Mass spectrometry for organic chemists

Cambridge Chemistry Texts

Mass spectrometry for organic chemists

R. A. W. JOHNSTONE
Senior Lecturer in Organic Chemistry
University of Liverpool

CAMBRIDGE
at the University Press 1972

Published by the Syndics of the Cambridge University Press
Bentley House, 200 Euston Road, London NW1 2DB
American Branch: 32 East 57th Street, New York, N.Y.10022

Library of Congress Catalogue Card Number: 72-75772

ISBNs: 0 521 08381 8 Hard covers
0 521 09685 5 Paperback

Printed in Great Britain
by William Clowes & Sons Limited
London, Colchester and Beccles

Contents

Conversion factors to Système Internationale (SI) units viii

Introduction ix

1 The mass spectrum 1
- 1.1. *Formation* 1
- 1.2. *Multiply charged ions* 6
- 1.3. *Isotopes* 8
- 1.4. *Metastable ions* 12
- 1.5. *Elemental compositions of ions* 12
- 1.6. *Appearance of the mass spectrum* 14

2 Apparatus 18
- 2.1. *Introduction* 18
- 2.2. *Inlet systems* 18
- 2.3. *Ion sources* 19
- 2.4. *Separation of ions* 21
- 2.5. *Recording of spectra* 24
- 2.6. *Miscellany* 25

3 Theory of mass spectrometry 26
- 3.1. *Introduction* 26
- 3.2. *Energy states resulting from ionization* 26
- 3.3. *Quasi-equilibrium theory (QET)* 36
- 3.4. *Qualitative theories* 39

4 Structure elucidation 46
- 4.1. *Classifications of mass spectra* 46
- 4.2. *Examination of the mass spectrum* 49
- 4.3. *Modification of mass spectra* 56
- 4.4. *Effects of structure on the appearance of the mass spectrum* 62
- 4.5. *Fragmentation of hydrocarbons* 64

4.6. *Primary fragmentations of aliphatic heteroatom compounds* 72
4.7. *Primary fragmentations of aromatic heteroatom compounds* 91
4.8. *Subsequent decomposition of primary fragment ions* 103
4.9. *Rearrangement accompanying fragmentation* 107
4.10. *Examples of structure elucidation by mass spectrometry* 114

5 Appendix 129
5.1. *Metastable ions* 129
5.2. *Accurate mass measurement* 138
5.3. *Combined gas chromatographic–mass spectrometric methods* 141
5.4. *Ionization and appearance potentials* 146
5.5. *Isotope analysis and labelling* 149
5.6. *Analysis of mixtures* 153
5.7. *Sample volatility and directed fragmentation* 156
5.8. *Ion cyclotron resonance spectroscopy* 157
5.9. *Chemical-ionization* 163
5.10. *Field-ionization* 166
5.11. *Relative abundances of isotope peaks* 168

Bibliography 171

Index 173

Dedicated to my Mother and Father,
and Christine, Steven, and Fiona.

Acknowledgements

The author thanks Mr F. A. Mellon and Dr S. D. Ward for helpful advice and discussion.

Conversion factors to SI units

1 kcal = 4.184 kilojoules (kJ)
1 eV = 1.602×10^{-22} kJ
1 Å = 10^{-10} metres (m)
1 torr = 133.3 newtons per sq. metre (N/m^2)

Introduction

> 'Why sometimes I've believed as many as six impossible things before breakfast.'
>
> The White Queen in *Through the Looking Glass*, Lewis Carroll (1832–98)

The basis of organic mass spectrometry is the production of ions from neutral species and an examination of the subsequent decomposition of those ions. Positive ions have been more thoroughly investigated since they are easier to produce and give more information through decomposition than negative ions.

Following the construction of the first mass spectrometers many years ago there followed a period of modest interest by physicists and chemists until it was realized by the organic chemist that here was an instrument which could be of immense value to his work. Mass spectrometry is only a more recent recruit to the range of physical methods used by organic chemists for elucidation of molecular structures but its use has increased rapidly and it has the advantage that only small quantities of material are required; the disadvantage is that the material is not recoverable i.e. it is not really a physical technique but a chemical reaction.

In his own inimitable fashion, the organic chemist has been able to extract a great deal of information after introducing empirical rules and concepts to mass spectrometry, and the technique is now an extremely useful analytical tool. At this stage it is worth reminding the reader that mass spectrometry alone seldom gives a unique solution to a problem and it is best used carefully in conjunction with information obtained by other means. The empirical rules and concepts so useful for aiding the interpretation of mass spectra to elucidate molecular structures have frequently been stretched by the organic chemist in attempts to explain 'mechanisms' of ion fragmentation. Sound physico-chemical principles have often seemed to be afterthoughts when welded into organic chemists' theories of mass spectrometry, but there are now signs of a more unified approach. It must be stressed that even without any sound basic understanding of the mechanisms of mass spectrometric fragmentations, the accumulated empirical rules allow mass spectrometry to be used sensibly for analytical purposes.

Chemical reactions in solids, liquids, and the gas phase are usually discussed in terms of isolated molecules whereas the actual behaviour is the result of 'group' effects arising from collisional activation of molecules. Mass spectra (except chemical-ionization mass spectra) are measured at low pressures when ion–molecule reactions are rare so that their interpretation in terms of isolated species is more satisfactory. However, for practical purposes these low concentrations of ions impose severe restraints on investigating them. The concentrations of ions are so low that other common physical methods such as ultraviolet, infrared, and nuclear magnetic resonance spectroscopy cannot be used to elucidate their structures and internal energies. At best, postulated ion structures can only be described as consistent with indirect evidence based on isotopic labelling, thermochemical arguments, and reasoning by analogy. As its name implies, a mass spectrometer measures mass (or strictly mass-to-charge ratio) and gives no direct information on ion structures. Because of this lack of knowledge of structures, fundamental investigation of the mechanisms of ion decomposition has been severely hampered.

To write a simple introductory text to this subject is a somewhat uncomfortable task since it becomes necessary to 'cut corners' and make statements which may be arguable. In this introduction to the subject, it has been felt necessary to include some theoretical aspects of mass spectrometry as well as its empirical applications, because by understanding fundamentals an empiricist may make better use of the method. The first part of the book covers some of the theoretical background but includes only a short section on apparatus since the latter is adequately dealt with in many other texts. The second part of the book deals with the empirical application of mass spectrometry to structure elucidation. Throughout, the various aspects of mass spectrometry are introduced as required and briefly explained, but where more detailed explanations are desirable, these have been included in the appendix. Finally, a bibliography is included of the more important texts on the subject which the reader is advised to consult for further information in conjunction with the very large number of research publications.

It was difficult to decide whether quantities should be described in the familiar but non-uniform units system (kcal, eV, Å, etc.) or in SI units. Since only a tiny proportion of data is as yet presented in SI units and these latter have not been universally accepted, the non-uniform system has been retained. Simple conversion factors appear at the front of the book and, in some tables, quantities are given in both systems for the convenience of the reader.

1 The mass spectrum

1.1 Formation. Most research into the mass spectrometry of organic compounds has been carried out on positive ions and these are discussed in detail. Negative ions and ion-pair formation are mentioned as the occasion arises.

When a molecule is ionized, a *molecular ion* ($M^{+\cdot}$) is produced and this may contain sufficient internal energy to fragment by ejection of a neutral particle (m^0) and the formation of a *fragment ion* (A^+). A neutral molecule gives a radical–cation as the molecular ion, and the fragment ion may be a cation or a radical–cation. The ejected neutral particle (m^0) may be a radical or neutral molecule.

$$M \xrightarrow{-e} M^{+\cdot} \longrightarrow A^+ + m^0$$

If the fragment ion (A^+) has sufficient internal energy, then further decomposition may occur with the formation of new fragment ions (B^+, C^+, etc.) until there is insufficient internal energy in any one ion for further reaction.

$$M^{+\cdot} \xrightarrow{-m^0} A^+ \xrightarrow{-m^0_a} B^+ \xrightarrow{-m^0_b} C^+ \xrightarrow{-m^0_c} \text{etc.}$$

Such a series of decompositions when elucidated from a mass spectrum is a *fragmentation pathway*. The molecular ion ($M^{+\cdot}$) and any of the fragment ions (A^+, B^+, C^+, etc.) may decompose by more than one pathway. The various fragmentation pathways together compose a *fragmentation pattern* characteristic of the compound under investigation. At one extreme, the fragmentation pattern might consist of only one pathway and result in a very simple mass spectrum. At the other extreme, the fragmentation pattern contains many, often interlocking pathways producing a complex spectrum. The extent to which fragmentation takes place along the individual pathways is determined by the amount of internal energy originally imparted to the molecular ion ($M^{+\cdot}$) and by the time allowed between ion formation and collection. Hence, the *mass spectrum* is not simply the fragmentation pattern but is the appearance of the fragmentation pattern at specified energies and times.

A mass spectrometer is designed to separate and measure the masses of molecular and fragment ions by making use of their mass-to-charge (m/e) ratios. An ion is usually formed with a single positive charge ($e = 1$) so that m/e is then equivalent to m and gives the mass of the ion directly.

During any one interval of time, molecular ions will be produced with various internal energies; each molecular ion will fragment at a rate determined by its initial energy because, at the low pressures normally obtaining in an ion source, ion–ion and ion–molecule collisions are rare so that collisional equilibration of internal energies does not occur (except in chemical-ionization – see later, p. 21). Some molecular ions may have insufficient energy to fragment whereas others have so much energy that decomposition proceeds right through the fragmentation pathway. Because of the initial range of internal energies in the molecular ions, a short period of time after ionization would see the presence of ions M^+, A^+, B^+, C^+, etc. in amounts determined by their individual rates of formation and decomposition and the initial energy imparted to M. If a sample of these ions in the ion source is obtained and their relative amounts measured, the results can be displayed with m/e values as abscissae and ion abundances as ordinates (figure 1.1). Such a picture of a sample of the ions in the

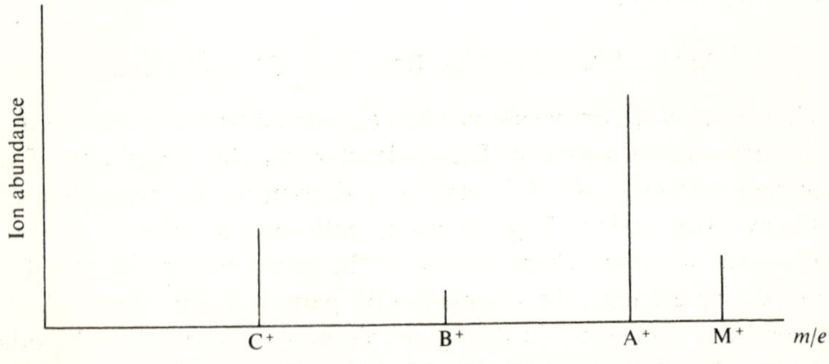

Figure 1.1. Abundances of ions M^+, A^+, B^+, C^+ in sample from ion source.

ion source is the basis of the mass spectrum which may be recorded on a cathode ray tube, paper chart, or photographic plate. An example of part of a mass spectrum recorded as peaks on a paper chart is shown in figure 1.2. As a standard practice, it is usual to make a record of a

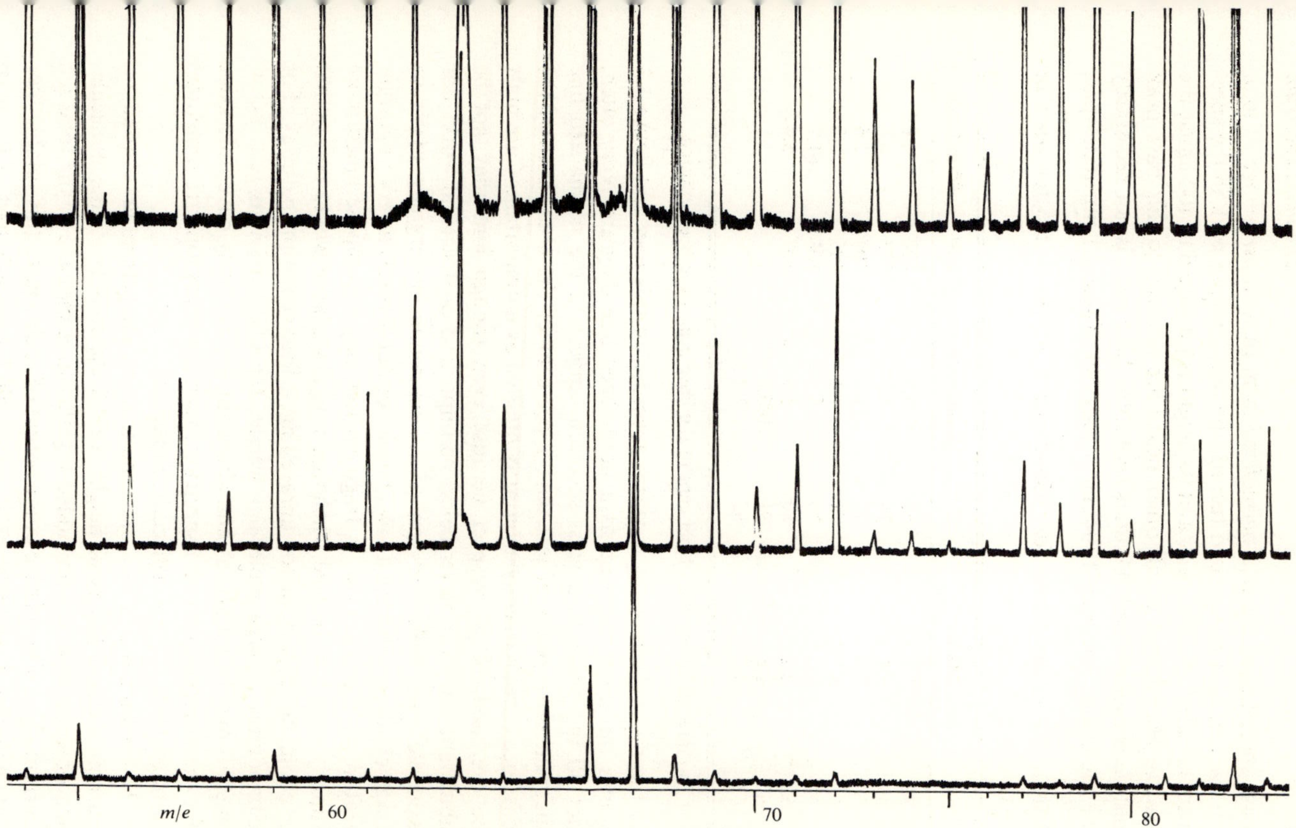

Figure 1.2. Part of a mass spectrum recorded on photographic paper (the three traces record the spectrum at increasing sensitivities from the lower to the upper trace).

mass spectrum in either a *normalized* form or as a *percentage of total ion current*. In a normalized record, the most intense peak in the spectrum is called the *base peak* and its height is put equal to 100 units; the relative abundances of all other peaks are referred to this base peak and lie between 0 and 100 units (note: the base peak is not necessarily the molecular ion peak although it may be). Table 1.1 illustrates part

TABLE 1.1 *Normalized ion abundances† in the mass spectrum of m-xylene*

m/e	Abundance	*m/e*	Abundance	*m/e*	Abundance
38	1.2	65	5.5	102	1.1
39	7.1	74	1.1	103	6.1
50	2.8	77	11.3	104	2.6
51	9.1	78	5.1	105	28.7
52	5.0	79	6.5	106	61.7
53	2.2	89	1.9	107	5.4
62	1.6	91	100.0		
63	4.2	92	7.8		

† Ion intensities less than 1 per cent of the base peak have been omitted.

of the mass spectrum of *m*-xylene recorded in this way and figure 1.3 shows the same spectrum as a normalized line diagram. Very often, not all the peaks in a mass spectrum appear in the normalized line diagram since peaks of less than 1 per cent of the abundance of the base peak are frequently arbitrarily omitted as unimportant. Care must be exercised in this respect because even though ions may not be abundant, they can nevertheless be important in structural work. For example, the molecular ion of a compound undergoing extensive fragmentation will have a low abundance but it is still a very important feature of the spectrum. A fragment ion may decompose much more rapidly than it is formed when again its abundance would be low but it could be important for unravelling a fragmentation pathway. For compounds of high molecular weight, the low mass ions (e.g. below *m/e* 50) may be numerous and abundant but also may have little value for the interpretation of a spectrum; such low mass ions are sometimes omitted from the normalized line diagram.

To obtain a record of a mass spectrum by the percentage of total ion current method the peak heights of all the ions from the molecular ion down to $m/e = n$ are added together as a measure of the total ion current (Σ_n). The relative contribution of each ion to this total is then calculated as a percentage ($\% \Sigma_n$). Table 1.2 records part of the mass

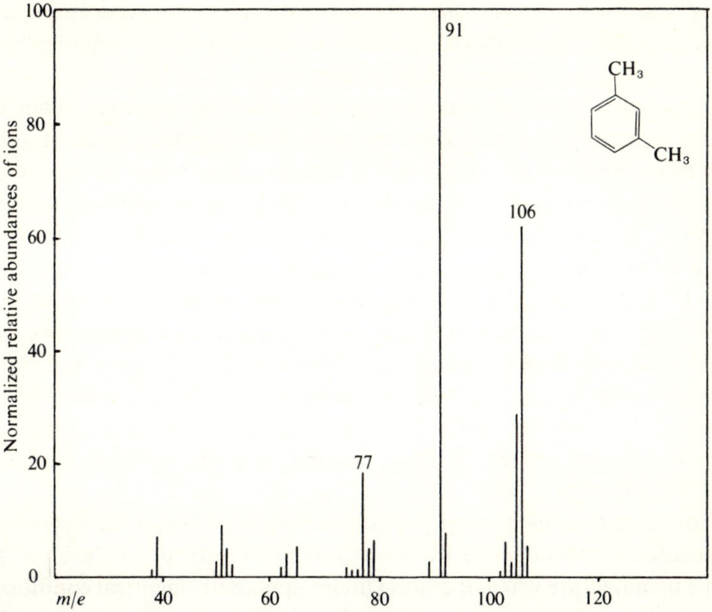

Figure 1.3. Line diagram showing normalized relative abundances of ions in the mass spectrum of *m*-xylene; *m/e* 91 = 100 units.

TABLE 1.2 *Ion abundances*† *in the mass spectrum of m-xylene as* $\% \Sigma_{38}$

m/e	Abundance	*m/e*	Abundance	*m/e*	Abundance
39	2.5	65	1.9	92	2.7
50	1.0	77	3.9	103	2.1
51	3.2	78	1.8	105	10.0
52	1.7	79	2.2	106	21.4
63	1.5	91	34.8	107	1.9

† Ion intensities of less than 1 per cent are omitted.

spectrum of *m*-xylene for $n = 38$ with abundances of ions as $\% \Sigma_{38}$. This last method of recording a mass spectrum is used less frequently than the normalization method but has some advantages in emphasizing the relative importance of an ion in the whole spectrum. The percentage ion current method finds some favour in comparing the spectra of isomers when changes in the relative contributions of certain ions to the total ion current may be sufficient to distinguish between the isomers. The difference between the two ways of recording spectra in tabular form is more apparent than real since the relative abundance of a peak in a normalized spectrum is simply related numerically to its value as a percentage of total ion current.

Two main facets of mass spectrometry will be apparent from the above description: (i) In contrast with what happens in most other common methods of physico-chemical analysis, some or all of the sample is consumed in mass spectrometry and is not recoverable. However, mass spectrometers are very sensitive instruments and spectra may be obtained from a few nano- or microgrammes of material. (ii) Unlike other physical methods, mass spectrometry does not deal with a well-defined property of a molecule. The appearance of a mass spectrum depends upon the interval of time between ionization and collection of ions, upon the initial energy distribution in the molecular ions and therefore on the method of ionization, and partly upon the physical characteristics of the instrument and therefore on its design and manufacture.

For these reasons it is not possible to guarantee that a mass spectrum is accurately reproducible from instrument to instrument – even of the same manufacture when operating under apparently identical conditions. Normally, variations in ion abundance of a few per cent are acceptable in comparing spectra from different laboratories, but generally, good uniformity is found and not too much difficulty is experienced from small inconsistencies.

1.2. Multiply charged ions. In organic mass spectrometry, ions are generally produced with single positive charges but may sometimes have two, three, or more charges so that the corresponding mass-to-charge ratios correspond to half, one-third, or lower fractional masses. A molecular ion with a single positive charge corresponds to a mass m, but with two positive charges it corresponds to $m/2$. Doubly charged ions are fairly common, especially in the mass spectra of aromatic compounds, but more highly charged ions are rarer. Figure 1.4 shows part of a mass spectrum containing some prominent doubly charged ions amongst the singly charged ones. The doubly charged species may fragment either into two singly charged ions or into another doubly charged ion with ejection of a neutral particle.

$$M^{++} \nearrow A^{+} + B^{+}$$
$$M^{++} \searrow C^{++} + m^{0}$$

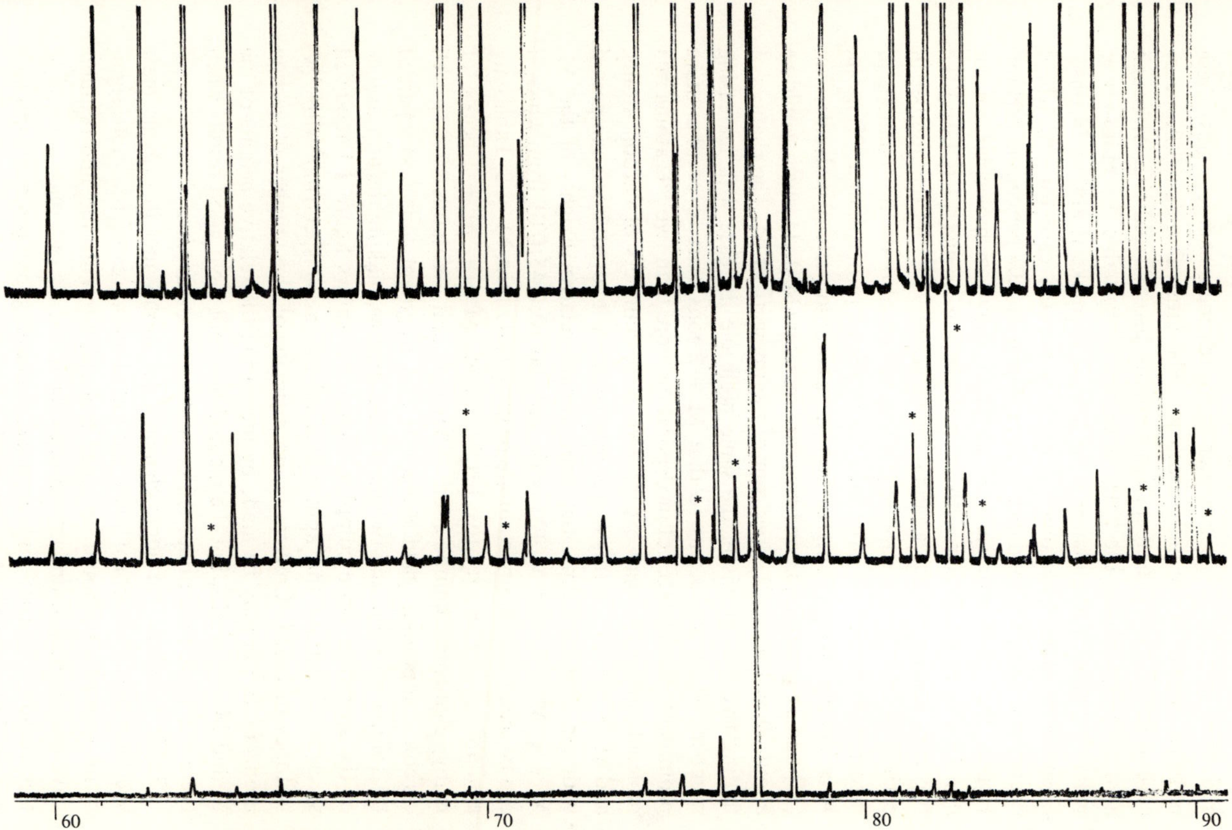

Figure 1.4. Partial mass spectrum showing doubly charged ions marked with an asterisk on the middle trace.

1.3. Isotopes. Many elements in their natural states contain isotopes, and because a mass spectrometer measures mass/charge ratios, these isotopes appear in the mass spectrum. The most abundant isotope of carbon is ^{12}C, but natural carbon contains also ^{13}C and ^{14}C. Although as a beta-ray emitter the latter is extremely valuable for radio-tracer work, its natural abundance is so low as to make it almost inconsequential to mass spectrometry. Such is not the case for ^{13}C which occurs in a natural abundance of approximately 1.08 per cent in carbon. Hence, the mass spectrum of methane shows a molecular ion at m/e 16 ($^{12}CH_4$) together with an isotopic ion at m/e 17 ($^{13}CH_4$) the two ions having relative abundances of about 99:1. As the number of carbon atoms in a compound increases, so also do the chances of incorporating one ^{13}C atom into the molecule rather than a ^{12}C atom. A compound with twenty carbon atoms would yield a molecular ion (M) and an isotopic ion one mass unit greater (M + 1) which would be about $20 \times 1.08 = 21.6$ per cent of the intensity of M. The chances of finding two ^{13}C atoms in the same molecule increase with increasing numbers of carbon atoms so that M + 2 ions start to become more prominent. The chances of incorporating more than two ^{13}C atoms in a molecule are very small except with large numbers of carbon atoms, and M + 3 ions from this source may be safely ignored. For a compound with ten carbon atoms, the approximate relative heights of the M, M + 1, M + 2, and M + 3 peaks are 100:10:0.45:0.01 from which it can be seen how unimportant are peaks greater than M + 2. Tables are available giving the relative heights of M, M + 1, M + 2 peaks for many elemental compositions (Beynon and Williams, 1963). All carbon compounds yield molecular and fragment ions accompanied by isotopic ions one and two mass units greater, making the appearance of the spectrum more complex. Figure 1.5 shows figure 1.1 modified so as to illustrate the appearance of the mass spectrum with ^{13}C isotopes included. Conversely, an estimate of the number of carbon atoms in an ion may be made by measuring the relative heights of the M and M + 1 peaks, but the method does not yield good results when there are more than about ten or twelve carbon atoms.

Table 1.3 gives, for some elements commonly met in organic mass spectrometry, the approximate natural abundances of the more significant isotopes. It should be noted that sulphur, chlorine, and bromine have particularly abundant isotopes separated by two mass units. For this reason, chlorine and bromine containing compounds especially are readily recognized in mass spectrometry and, by examining the

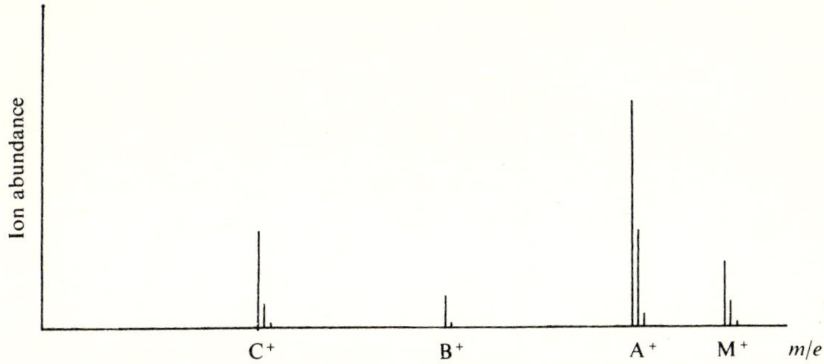

Figure 1.5. The simple spectrum shown in figure 1.1 with the addition of ^{13}C-isotopic masses for ions M^+, A^+, B^+, C^+.

TABLE 1.3 *The more important natural isotope abundances for elements commonly occurring in organic mass spectrometry*†

Element	Isotope (% natural abundance)‡		
Hydrogen	^{1}H (99.9$_9$)		
Carbon	^{12}C (98.9)	^{13}C (1.1)	
Nitrogen	^{14}N (99.6)	^{15}N (0.4)	
Oxygen	^{16}O (99.8)	^{18}O (0.2)	
Fluorine	^{19}F (100.0)		
Silicon	^{28}Si (92.2)	^{29}Si (4.7)	^{30}Si (3.1)
Phosphorus	^{31}P (100.0)		
Sulphur	^{32}S (95.0)	^{33}S (0.7)	^{34}S (4.2)
Chlorine	^{35}Cl (75.5)	^{37}Cl (24.5)	
Bromine	^{79}Br (50.5)	^{81}Br (49.5)	
Iodine	^{127}I (100.0)		

† Metals frequently possess many abundant isotopes and, because of the diversity of organometallic compounds, a listing of metal isotopes has not been included here. Many reference texts contain listings of isotope abundances; see also Beynon (1960), Kiser (1965).

‡ Percentages are given correct to the first decimal place. Trace isotopes are not included as they have little consequence in mass spectrometry at their natural abundance levels.

isotope pattern in the molecular ion region, the numbers of chlorine and bromine atoms in the original molecule may be determined (Beynon, 1960; Biemann, 1962). Figure 1.6 shows the appearance of the molecular ion region for a dichlorobenzene; note the major peaks spaced two

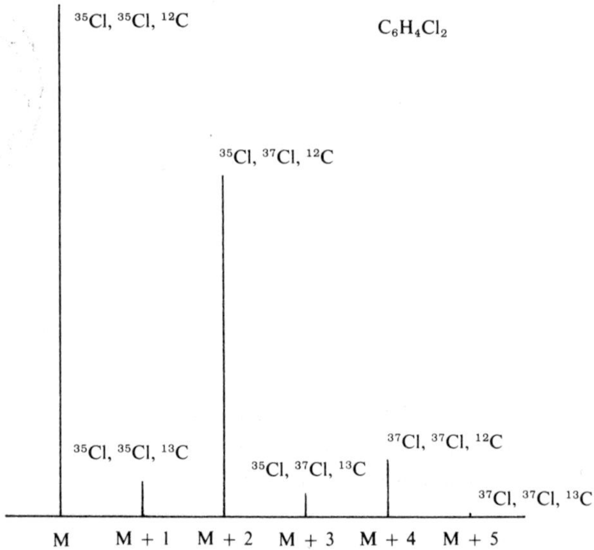

Figure 1.6. Contributions of carbon and chlorine isotopes to the pattern of peaks in the molecular ion region of a dichlorobenzene. The molecular ion (M) is usually considered to be the peak having contributions from the lowest mass isotopes but strictly all six ions are molecular ions.

mass units apart due to the ^{35}Cl and ^{37}Cl isotopes of the chlorine atoms and the minor peaks caused by contributions from the ^{13}C isotope. A method of calculating simple isotopic patterns is presented in §5.11.

Normalized mass spectra are sometimes drawn with the isotope contributions simplified. For example, for two peaks at m/e 253, 254 of relative abundances as shown in figure 1.7*a*, the ^{13}C isotope contribution of m/e 253 contributes to the height of the peak at m/e 254; the ^{13}C isotope of m/e 254 occurs at m/e 255. If there were eighteen carbon atoms in the elemental composition of the ion at m/e 253, then approximately $18 \times 1.1 = 19.8$ per cent of its peak height is the ^{13}C contribution in m/e 254. By subtracting this ^{13}C contribution from m/e 254 the relative abundances of the ions appear as in figure 1.7*b*; similarly the ^{13}C isotope peak of m/e 254 at m/e 255 disappears. This device of removing isotope contributions is extremely helpful for many organometallic compounds since metals frequently have several isotopes and so yield complex-looking spectra e.g. mercury has seven isotopes spread over nine mass units. Each metal atom containing ion can

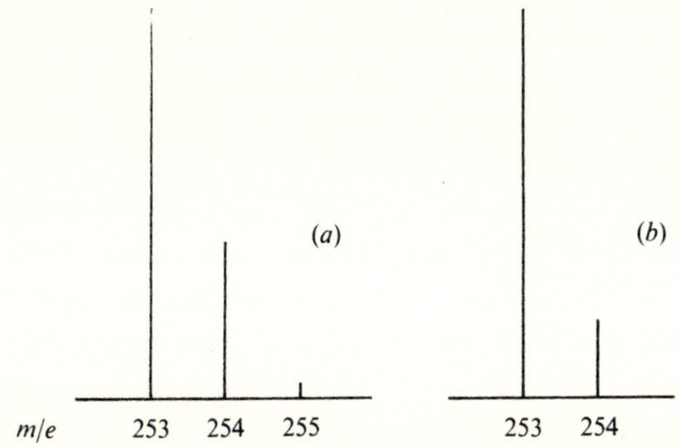

Figure 1.7. Appearance of part of the mass spectrum of a compound with 18 carbon atoms at *m/e* 253–255 (*a*) with ^{12}C and ^{13}C isotope contributions; (*b*) with ^{13}C isotope contribution removed and therefore showing only ^{12}C isotopes.

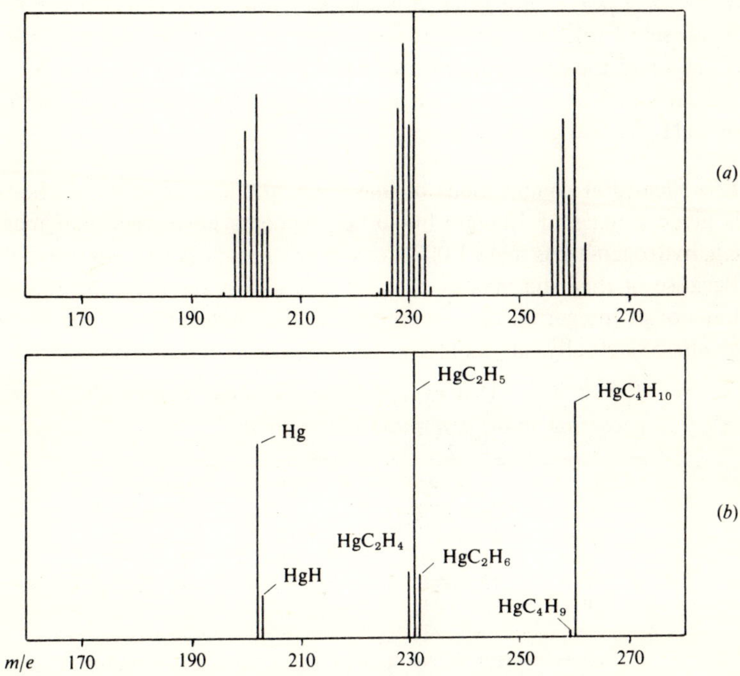

Figure 1.8. Partial mass spectrum of diethylmercury, (*a*) including all isotope peaks, and (*b*) monoisotopic spectrum based on ^{202}Hg and ^{12}C.

therefore afford many isotope peaks and removing their separate contributions simplifies the interpretation of such a spectrum. Figure 1.8 shows the mass spectrum of diethylmercury both with and without the isotope contributions; the spectrum without isotope peaks is much easier to interpret.

1.4. Metastable ions. The sharp peaks of the normal ions in a mass spectrum are usually accompanied by some much broader low intensity peaks due to the so-called *metastable ions.* The maxima of these metastable ion peaks normally occur at non-integral m/e values and are easily distinguished from normal ions (figure 1.9). The peak shapes are approximately Gaussian but occasionally flat-topped ones are found extending over several mass units. The origin of the metastable ions is discussed later. These ions are very useful for determining fragmentation pathways; if an ion of mass m_1 decomposes to give an ion of mass m_2, then a metastable ion may be found at mass $m^* = m_2^2/m_1$. Observation of a metastable ion therefore at this position (m^*) confirms the fragmentation of a *parent* ion (m_1) to a *daughter* ion (m_2). Thus, in the mass spectrum of toluene abundant ions occur at m/e 91 (C_7H_7) and m/e 65 (C_5H_5) and the appearance of a metastable ion at m/e 46.4 ($= 65^2/91$) indicates that at least some of the ions at m/e 65 arise through ejection of C_2H_2 from the ion at m/e 91.

1.5. Elemental compositions of ions. On the atomic scale, carbon-12 is given a mass of 12.0000 but other elements have fractional masses e.g. hydrogen-1 has mass 1.0078, oxygen-16 has mass 15.9949 (table 1.4). Because of this, the masses of most ions in a mass spectrum fall near but not at integer values. For instance, the molecular ion of acetone at m/e 58 actually has a mass of 58.0418.

TABLE 1.4 *Masses of some elements commonly occurring in organic mass spectrometry*†

^{1}H	1.00782	^{32}S	31.97207
^{12}C	12.00000	^{35}Cl	34.96885
^{14}N	14.00307	^{37}Cl	36.96590
^{16}O	15.99491	^{79}Br	78.91839
^{19}F	18.99840	^{81}Br	80.91642
^{31}P	30.97376		

† Given correct to five decimal places. Notice how on this atomic scale, with ^{12}C taken as standard, the masses fall either just above or just below integral values.

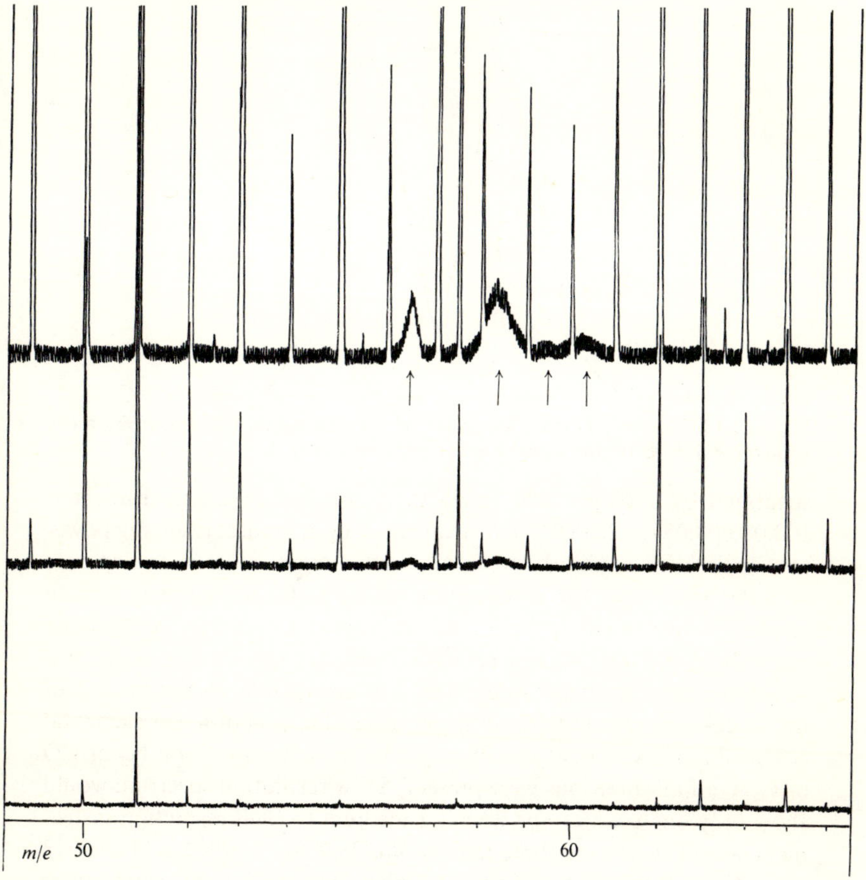

Figure 1.9. Metastable ion peaks indicated by arrows in a mass spectrum (normal ions give narrow peaks at integral mass; metastable ions usually give broad peaks at non-integral mass). Note also the fairly abundant doubly charged ion at *m/e* 57.5.

The resolving power of a mass spectrometer is a measure of its ability to separate two ions of any defined mass difference. Basically, for two overlapping peaks M_1, M_2 (figure 1.10), the resolution may be defined in terms of the mass difference (ΔM) between them such that the peaks are said to be resolved if $(h/H) \times 100 \leqslant 10$, where H is the height of the peaks and h measures the depth of the 'valley' between them. The resolution is then the value of $M_1/\Delta M$ when $(h/H) \times 100$ is equal to 10. For example, when two masses (100.000, 100.005) are

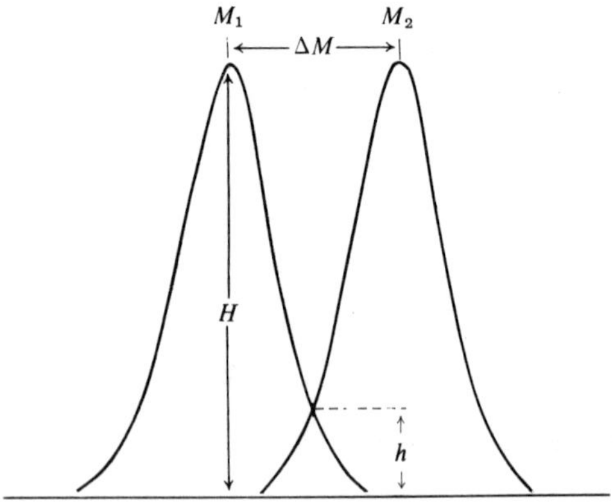

Figure 1.10. Two overlapping ion peaks M_1, M_2 of height (H) and overlap (h). If $(h/H) \times 100 \leqslant 10$, these peaks are considered resolved.

separated by a 10 per cent valley the resolution of the instrument is 100.000/0.005 i.e. 20 000. An instrument of even modest resolving power can distinguish readily between adjacent integral masses; to separate *m/e* 100 from *m/e* 101 a resolution of only 100 is required. A high resolution instrument, however, can separate an ion at *m/e* 100.000 from one at *m/e* 100.005. At low resolution, a mass spectrum consists of a series of peaks at integer *m/e* values but at medium resolution any of these peaks may be split due to the presence of ions of different elemental compositions. An ion at *m/e* 28 might have the composition N_2 or CO or C_2H_4; if all three ions were present, a low resolution spectrum would show only one peak at *m/e* 28 but a medium to high resolution instrument would reveal the three ions at *m/e* 28.0061, 27.9949, and 28.0313 respectively (figure 1.11). Thus, at higher resolution small mass differences may be detected and this property is utilized for accurate mass measurement of ions as described in §5.2.

1.6. Appearance of the mass spectrum. Beginning with the simple picture of ion decomposition, and introducing the complications of isotopes, multiply charged ions, metastable ions, and peak splitting due to differing elemental compositions of ions yields an overall impression of what to expect and what to look for in a mass spectrum. The mass spectrum of aniline in figure 1.12 has been chosen to illustrate these features. There is a further, practical complication evident in figure 1.12. Because of the great range in intensities of peaks from the low intensity metastable ions to the high intensity base peak, it is necessary to

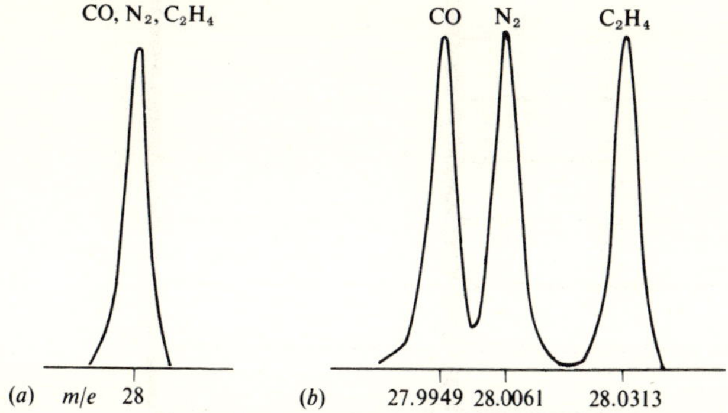

Figure 1.11. The appearance of the mass spectrum of a mixture of CO, N_2, and C_2H_4 at *m/e* 28 with (*a*) low resolving power showing no separation of masses, and (*b*) medium resolving power (3000) showing good separation.

be able to record the spectrum at different sensitivities simultaneously. Figure 1.12 shows three traces increasing in sensitivity from the bottom of the chart upwards. The lowest trace contains all the features of the *routine* or *low resolution spectrum*, the middle trace is used to aid the measurement of small peaks in the lower trace, and the upper trace is used both for counting the spectrum and looking for metastable ions. In counting a mass spectrum it is usual to first select some readily identifiable peaks of known *m/e* values. Air is usually present in or can be introduced into a mass spectrometer and provides a number of convenient 'marker' ions. Thus, peaks due to nitrogen (*m/e* 28), oxygen (*m/e* 32), and water (*m/e* 18) are observed, together with a doublet at *m/e* 40 due to argon and C_3H_4 from hydrocarbon 'background' ions. With very little experience, these ions are easily identified and used as starting points for counting up all the other ions in the spectrum. On a more sophisticated level, electromechanical mass markers may be purchased which put a mass count directly onto the spectrum as it is being recorded.

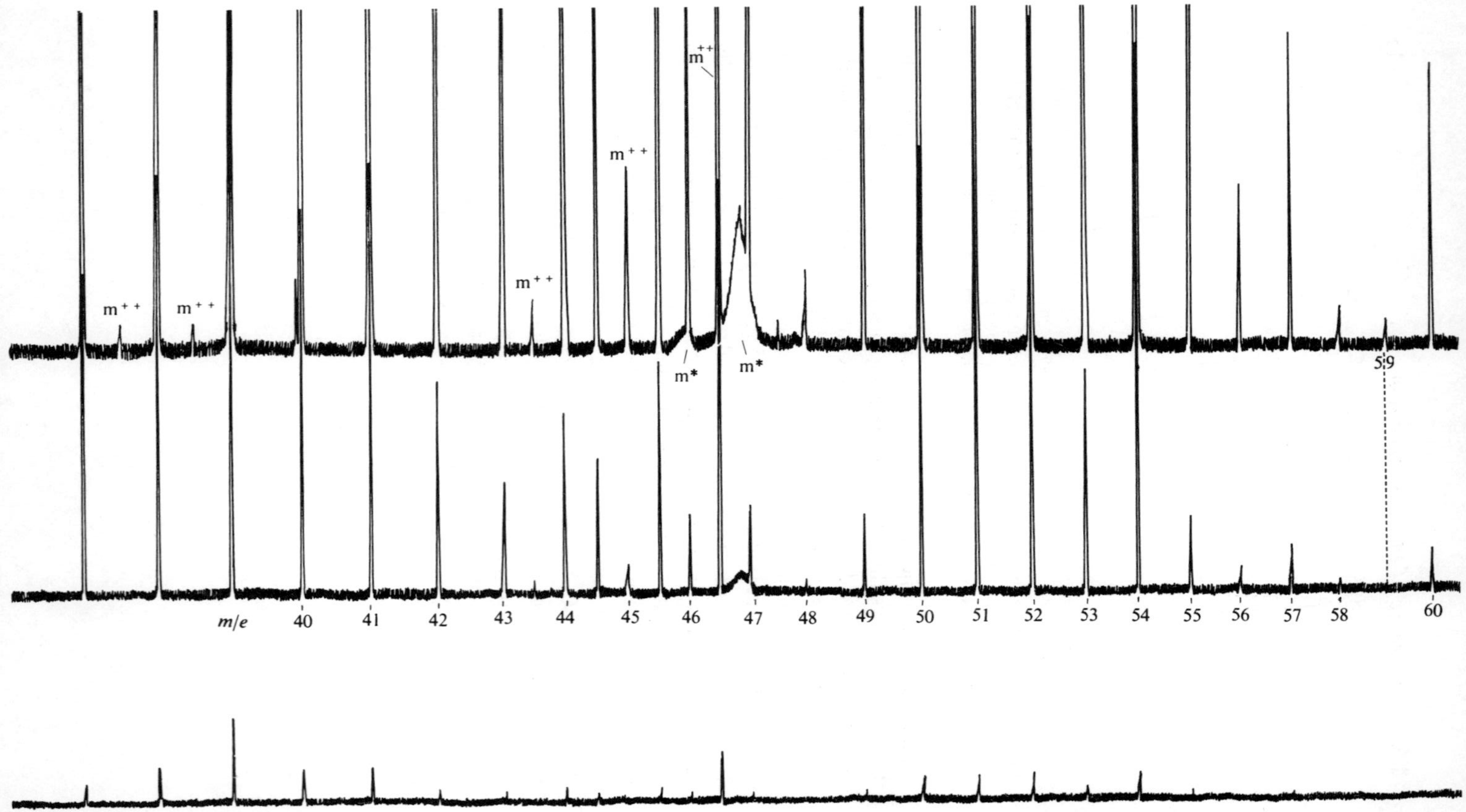
m⁺⁺
m⁺⁺
m⁺⁺
m⁺⁺
m⁺⁺
m*
m*
59
m/e
40
41
42
43
44
45
46
47
48
49
50
51
52
53
54
55
56
57
58
60

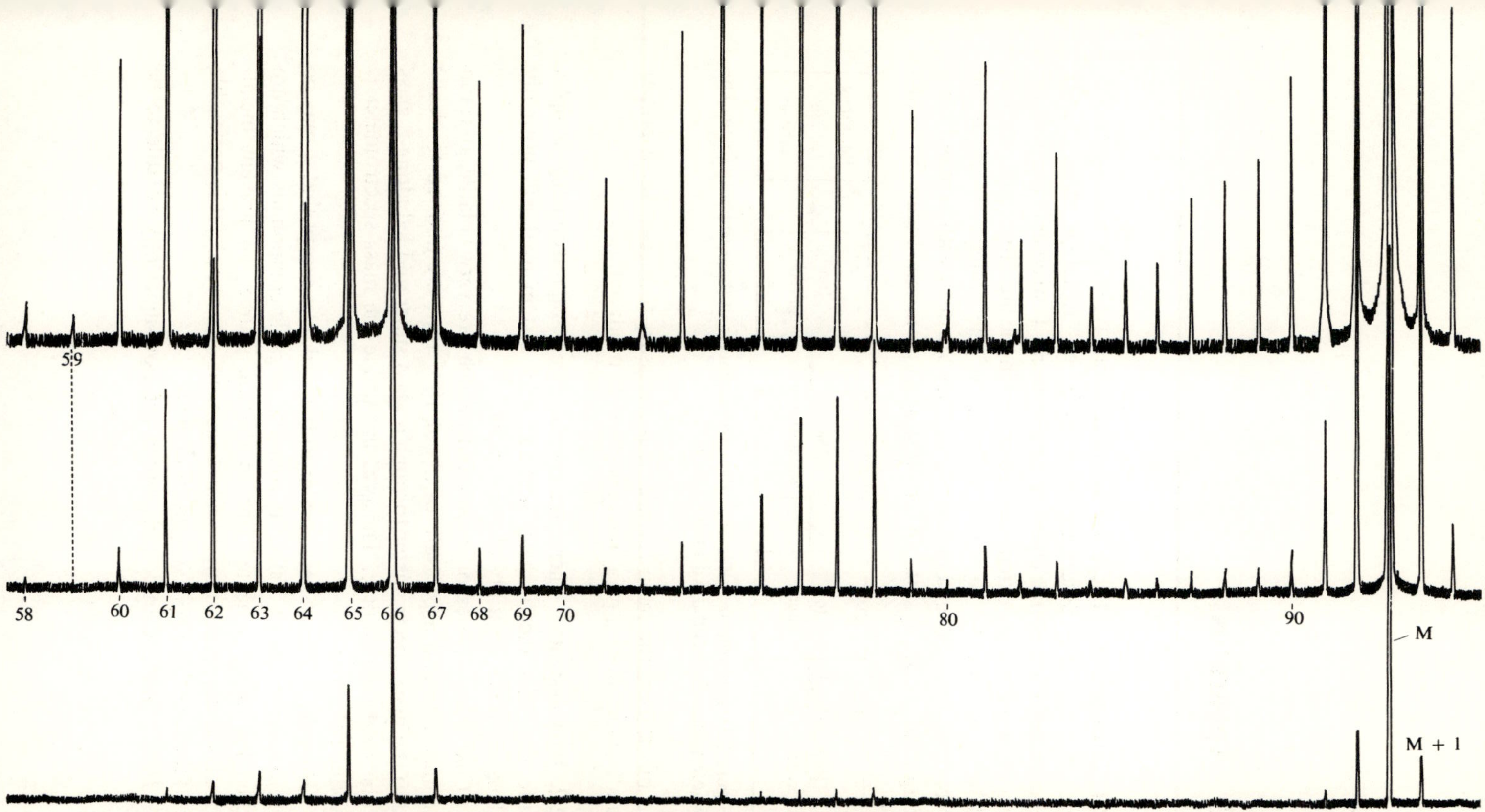

Figure 1.12. Mass spectrum of aniline. The molecular ion (M), and ^{13}C isotope ion M + 1 are shown. Metastable ions (m*) and doubly charged ions (m^{++}) are marked. Integral mass-to-charge ratios are given below the middle trace where the spectrum is counted; where a peak does not appear, as at *m/e* 59, a check on the top trace will often reveal a low abundance ion.

2 Apparatus

2.1. Introduction. There are many different types of mass spectrometer but the ones described here briefly are of a type most usually encountered in organic chemical applications. Instruments for measuring mass spectra are extensively described in the literature, and excellent descriptions are to be found in many of the volumes listed in the bibliography at the end of this book. They are dealt with here only in a diagrammatic form suitable for illustrating essentials. Figure 2.1 shows

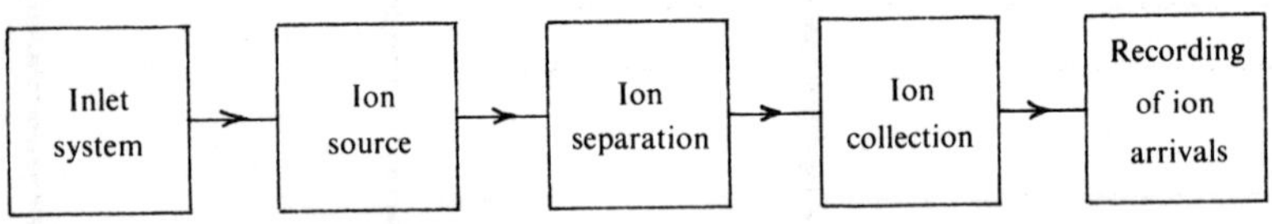

Figure 2.1. Essential features of a mass spectrometer in diagrammatic form. The whole system is maintained under vacuum.

in block form the main features it is necessary to discuss for a proper understanding of how a mass spectrum is produced.

2.2. Inlet systems. There are several means of introducing samples into the mass spectrometer and the inlet system used will normally depend on the volatility of the sample.

Cold inlet. Gases or compounds which are very volatile at room temperature and a pressure of about 10^{-2} torr, are allowed to 'leak' into the mass spectrometer through a glass sinter and led along a glass tube to the ion source. The pressure inside the mass spectrometer is about 10^{-6} torr.

Hot inlet. This is very similar to the cold inlet except that it may be heated up to about 300°C to volatilize organic compounds which are then led along a heated line to the ion source. To prevent metal-catalysed decomposition or rearrangement of a sample, it is preferable to fabricate the whole system from glass.

Direct inlet. Compounds which are not sufficiently volatile to be introduced through the hot or cold inlets may be inserted directly into the ion source by means of a probe passing through a vacuum lock. At the low pressures of about 10^{-6} torr inside the ion source and with heating, very many compounds are sufficiently volatile to yield good mass spectra. The system is so easy to use and requires so little material that it is frequently the method of choice even for relatively volatile substances. The probes are often heated by radiation from or contact with the ion source itself, but probes are now available which may be heated or cooled independently of the source. Using a direct insertion probe, compounds of molecular weight up to about 2000 may be measured fairly readily.

Gas chromatographic inlets. The effluents from gas chromatographic columns consist of carrier gas admixed with the compounds being investigated. Carrier gas flow rates of 50 ml/min are common in gas chromatography but such quantities of gas cannot be allowed directly into the ion source because large increases in pressure would result and lead to arcing of the high potentials inside the source. Consequently it is necessary to remove as much carrier gas as possible, ideally without diminishing at the same time the amount of sample going into the source. Differentiation between carrier gas and entrained sample can be effected by molecular separators (§5.3). A 100 per cent efficient molecular separator would remove all the carrier gas and pass all the sample into the ion source, but most separators are far from this efficiency.

2.3. Ion sources. An ion source may be simply defined as the region in which ionization occurs. As shown in figure 2.2, the region is usually

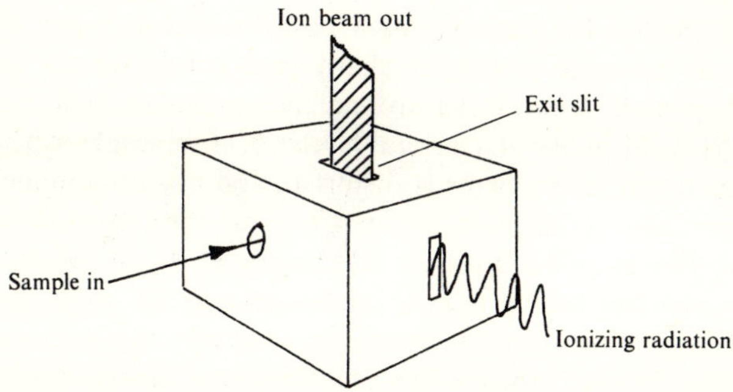

Figure 2.2. Essential features of an ion source for electron-impact or photo-ionization.

enclosed in an ion chamber in which the sample is ionized. The ions produced are pushed out towards an exit slit by a low positive potential inside the ion chamber and then accelerated through a high potential of 2–8 kV and passed away into the mass spectrometer for separation according to mass/charge ratio. There are several common means of ionizing organic compounds.

Photons. The ionization potentials of most organic compounds lie below 13 electronvolts so that high intensity, short wavelength radiation can be used for ionization. At energies greater than about 11 electronvolts there are no materials with which to make windows to transmit the radiation into the source. Therefore, it is necessary to arrange for the photon source to emit radiation directly into the ion source. The helium discharge at 21.21 electronvolts provides a convenient source of photons able to ionize all organic compounds. Unless a high intensity of radiation is attained inside the ion chamber low ion yields will be obtained with consequent low sensitivity of the mass spectrometer.

Electrons. Electrons, obtained from a heated filament *in vacuo*, are accelerated through some voltage (V) and directed across the ion chamber; the electrons thus have energy $e \cdot V$. The voltage (V) is continuously variable between about 5 and 100 volts and, by convention, standard mass spectra are obtained at 70 volts because maximum ion yield is obtained near this value. In practice, mass spectra may be obtained at any voltage down to the ionization potential of the compound under investigation, and it is often advantageous to reduce the energy of the impacting electrons much below 70 volts. In photon and electron-impact sources, the ions spend approximately 10^{-6} seconds after formation in the ion chamber before acceleration into the main part of the mass spectrometer.

Electric fields. By putting a high positive electric potential on an electrode in the shape of a point, sharp edge, or fine wire, a very high potential gradient is produced around the regions of high curvature. The electric field (molecular orbital levels) of a molecule experiencing such large potential gradients is distorted, and quantum tunnelling of an electron can occur from the molecule to the positively charged electrode; this is field-ionization. The positive ion so formed is immediately repelled by the positive electrode into the mass spectrometer. Although this is an effective way of producing ions, there are difficulties in focussing them into a beam. Because the initially produced ions are repelled by the high positive potential on the electrode immediately after formation, the residence time of these ions in the source region is short (10^{-12} seconds) compared with the residence time of ions in an electron-impact source (see §5.10).

Chemical-ionization. The collision of an ion and a molecule can lead to reaction giving a new charged species. Thus, the ion–molecule reaction between a methane ion and a methane molecule gives rise to the unusual and fairly stable CH_5^+ species.

$$CH_4^{+\cdot} + CH_4 \longrightarrow CH_5^+ + CH_3^{\cdot}$$

For effective ion–molecule reaction it is necessary to operate at source pressures of about 10^{-4} torr, much greater than for other ion sources, and this leads to severe problems in preventing arcing of the high potentials in the source. It is also necessary to arrange the construction of the ion chamber in such a way that a pressure of 10^{-4} torr may be maintained within it whilst the pressure immediately outside it is 10^{-6} torr. At a pressure of 10^{-4} torr there are many ion–molecule collisions before the ions leave the ion chamber and collisional equilibration of energies may be expected. Because of this, the ions probably possess a distribution of internal energies similar to those found in solution chemistry. These so-called reactant gas ions (CH_5^+ above) can ionize other neutral organic molecules by proton transfer, or less usually, by hydride transfer. The technique has been termed chemical-ionization and is described more fully in §5.9.

$$AH^+ + M \longrightarrow A + MH^+$$

Combined sources. It is often an advantage to be able to generate mass spectra of organic materials by two different ionization methods. For this reason, combined electron-impact/field-ionization and electron-impact/chemical-ionization sources have been constructed to operate in either of the two possible modes. For example it is then possible to obtain the mass spectrum of an organic compound first by electron-impact ionization and then, by switching to the other mode, by field ionization. Differences between the two spectra may be very informative (see later sections).

2.4. Separation of ions. The separation of ions according to their mass-to-charge ratios can be achieved with magnetic and electric fields alone or combined. Most of the basic differences between the various types of mass spectrometers lie in the manner in which such fields are used to effect separation. It is not possible here to describe the many different ways in which the fields are combined, and so a description follows for a system typically used in organic mass spectrometry. Other systems are then briefly described.

After the beam of ions has been accelerated through a potential of 6000–8000 volts away from the ion source, it is divergent and slits are used to reduce the spread (figure 2.3). The ion beam then passes

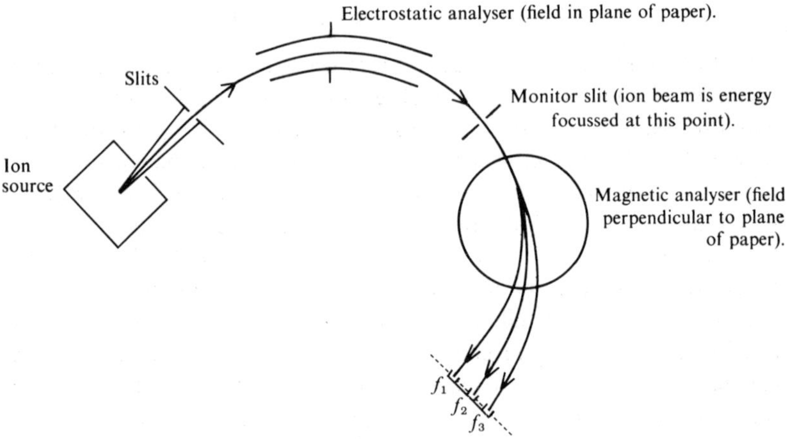

Figure 2.3. Flight paths of ions in a double focussing mass spectrometer of Nier–Johnson geometry.

between a pair of smooth, curved metal plates called an electrostatic analyser because the electric potential maintained across the plates focusses the ion beam at a second slit (monitor, figure 2.3). The electrostatic analyser (or electric sector) is an energy focussing device. Thus, if V is the potential through which the ions are initially accelerated and B is the field in the electric sector, then equation (2.1) is obtained where

$$r = 2.V/B \tag{2.1}$$

r is the radius of curvature of the ion path. The two equations governing the motion of the ions through the analyser are those relating their kinetic energy, $e.V = \frac{1}{2}.m.v^2$, and centrifugal force, $e.B = m.v^2/r$ in which v is the velocity of an ion of mass m and charge e; simple combination of these two equations gives equation (2.1). Thus ions accelerated through a potential (V) and passing through a uniform field (B) have the same radius of curvature (r) irrespective of mass-to-charge ratio. Therefore, if the field B is kept constant, the electric sector focusses the ions according to their translational energies.

The energy-focussed ion beam is then passed through a magnetic field where mass separation is effected. As described above, after acceleration from the ion source, the ions possess a translational energy $e.V(=\frac{1}{2}.m.v^2)$. On passing through the magnetic field, the ions experience a centrifugal force ($m.v^2/r$, where r is the radius of curvature) due to the constraint ($H.e.v$) imparted by the magnetic field (H). From these quantities, equation (2.2) readily follows.

$$\left[e.V = \tfrac{1}{2}.m.v^2;\ H.e.v = \frac{m.v^2}{r}\right] \qquad \therefore \frac{m}{e} = \frac{H^2.r^2}{2.V} \tag{2.2}$$

At constant values of the magnetic field (H) and the accelerating potential (V), each value of m/e will correspond to a radius of curvature r. Thus, after separation in the magnetic field, singly charged ions of mass m_1, m_2, m_3 will be brought to foci f_1, f_2, f_3 (see figure 2.3) corresponding to radii of curvature r_1, r_2, r_3. If an electrical detector is placed at f_1 then arrival of ions (m_1/e) will be recorded and if other detectors were placed at f_2, f_3 the ions m_2/e, m_3/e would also be recorded. Rather than have a series of detectors, it is much simpler to vary the magnetic field (H) continuously so that each m/e species is brought to a focus at f_1 in turn i.e. the radius of curvature (r) is kept constant; this is magnetic scanning of the spectrum. It is possible also to vary the electrical accelerating voltage (V), with H constant, to bring each m/e species to the same focus (see equation (2.2)); this method of voltage or electrical scanning is less desirable than magnetic scanning because the sensitivity of the instrument decreases with any decrease in the accelerating voltage.

The combined use of electric and magnetic analysers gives a *double focussing* mass spectrometer because the ion beam is focussed first for translational energy and then for mass-to-charge ratio. The layout of electric and magnetic fields described above (figure 2.3) is known as a *Nier–Johnson* geometry. If the electrostatic analyser is omitted from the instrument it becomes a *single focussing* mass spectrometer of lower resolving power. It is attainable resolving power which is the important difference between single and double focussing instruments. A good single focussing mass spectrometer can distinguish integer m/e values up to masses well above 1500; a good double focussing instrument can distinguish individual elemental compositions at each m/e value in the same range. Because of the greater resolution obtainable, double focussing machines are used for accurate mass measurements on ions.

Other layouts of electrical and magnetic fields are met in other types of mass spectrometer, but the basic aim remains the separation of ions according to m/e value with as great a resolution as is needed. Instruments using the *Mattauch–Herzog* geometry have their fields arranged as in figure 2.4. Electrical detection of ions is not always used and

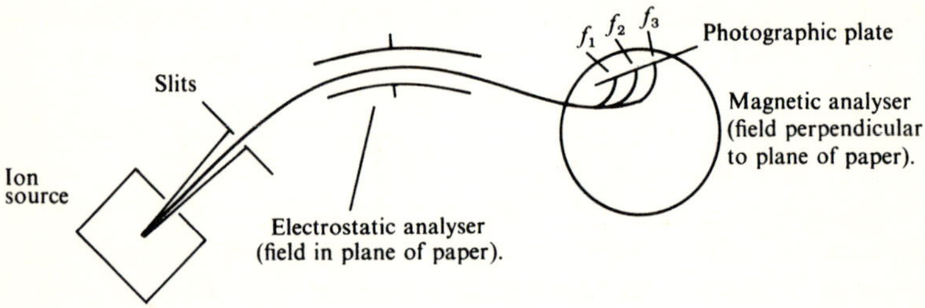

Figure 2.4. Flight paths of ions in a double focussing mass spectrometer of Mattauch–Herzog geometry.

instead a photographic plate may be placed in the plane of the ion foci f_1, f_2, f_3 (figure 2.4) and the magnetic field kept constant. The ions then give rise to narrow bands on the photographic plate and the density of these bands is a measure of the number of ions arriving. The sensitivity range of a photo-plate is limited and it is usually necessary to make several exposures for increasing periods of time in order to record adequately all ions from the most to the least abundant. The photo-plate method of detection is clumsier than the electrical since after each spectrum the plate must be developed. Also, the density of the bands must be found with a second instrument, a microdensitometer, which scans the bands and gives out optical density readings as electrical signals (see §5.2, figure 5.9). One big advantage of the photo-plate is that fluctuations in ion intensities are immaterial since all ions are recorded simultaneously whereas, with electrical detection and magnetic scanning of the spectrum, fluctuations in ion intensities during the scan result in an alteration in the appearance of the spectrum.

2.5. Recording of spectra. The electrical signals from the detector are amplified and passed to a recording device. A pen recorder may be used but this means the spectrum must be scanned rather slowly because of the relatively slow response time of the pen mechanism. Alternatively, the amplified signals are used to deflect mirror galvanometers and ultraviolet light reflected from them is focussed onto sensitized paper to give a permanent record. The response times of such galvanometers are very short and spectra may be recorded very quickly.

At a more sophisticated level, the electrical signals are passed, either directly, or via a magnetic tape recording, to a computer which then evaluates the incoming data and prints out the required information in the form of a normalized spectrum or a listing of the accurate masses of all the ions in the spectrum. The photographic plate method of detection of ions provides its own record of the spectrum, and the electrical signals from the microdensitometer scan of the plate may be passed to a computer also.

2.6. Miscellany. A brief mention of some other types of mass spectrometer less widely used for organochemical applications follows.

The *time of flight* instrument separates ions of different masses by making use of their different velocities after acceleration through a potential (V); no magnetic sector is used. Since $e.V = \frac{1}{2}.m.v^2$, the velocity (v) of an ion of mass (m) is $v = (2.e.V/m)^{1/2}$ i.e. the velocity is mass dependent. Thus, if a bunch of ions (m_1, m_2, m_3, etc.) is accelerated and then allowed to pass along a field-free region, the ions will arrive at a detector at different time intervals depending on their velocities (v_1, v_2, v_3, etc.). These time intervals are very short and it is necessary to display the mass spectrum on a cathode ray tube. Generally, the resolving power of these instruments is low but because of their fast response times, they find considerable use in investigations of fast reactions.

The *ion cyclotron resonance* mass spectrometer employs crossed electric and magnetic fields. This type of instrument has found increasing use recently particularly for studying ion–molecule reactions and is described more fully in §5.8.

The *quadrupole* mass spectrometer consists essentially of four electrodes between which the ions pass. By suitably arranging radiofrequency fields between the electrodes any mass may be selected and passed through the system, but passage of all other ions of different mass is prevented. Because of this filtering action the quadrupole is often called a mass filter. Generally of modest resolving power, the quadrupole has advantages of small size, robustness, and relative cheapness.

Finally, many of the books listed in the bibliography contain detailed descriptions of all types of mass spectrometers and should be consulted for further information.

3 Theory of mass spectrometry

3.1. Introduction. A thorough understanding of mass spectrometric processes is severely hampered through lack of knowledge of ion structures and their electronic states. The fragmentation of ions depends on the excess of internal energy they contain and this is governed mostly by the initial ionization step. Despite the basically poor knowledge of ion structures, many generalizations may be made which are useful for understanding mass spectrometric fragmentation of organic compounds.

3.2. Energy states resulting from ionization. Consider a diatomic molecule (M) ionized to yield a molecular ion ($M^{+\cdot}$). The vibrational energies of both the molecule and the molecular ion may be represented in the usual way by Morse curves as shown in figure 3.1. For the moment, it is only necessary to consider the zeroth vibrational energy level for M because even at 500°K there is little population of the higher vibrational levels. However, it is necessary to consider the vibrational energy levels, $\nu = 0, 1, 2, \ldots$ in the ion ($M^{+\cdot}$). If ionization is very fast, then by the Franck–Condon principle, there will be a vertical transition between M and $M^{+\cdot}$. The minima of the two Morse curves for M and $M^{+\cdot}$ are unlikely to coincide so that a vertical transition will leave $M^{+\cdot}$ in a vibrational state above the zeroth level. Transition *a* in figure 3.1. for example gives the ion in the vibrational state $\nu = 4$ and the ion therefore has an excess of internal energy. A vertical transition may occur to a vibrational level in $M^{+\cdot}$ above the dissociation limit and in this case the ion will contain energy in excess of the dissociation energy (*D*). Accordingly, $M^{+\cdot}$ will fragment as soon as it begins to vibrate. It is also possible that the ground state of the ion $M^{+\cdot}$ is completely dissociative (transition *b* figure 3.1.) so that decomposition of the ion follows immediately after its formation.

From a number of molecules (M), ions ($M^{+\cdot}$) are produced with different amounts of internal energy proportional to the transition

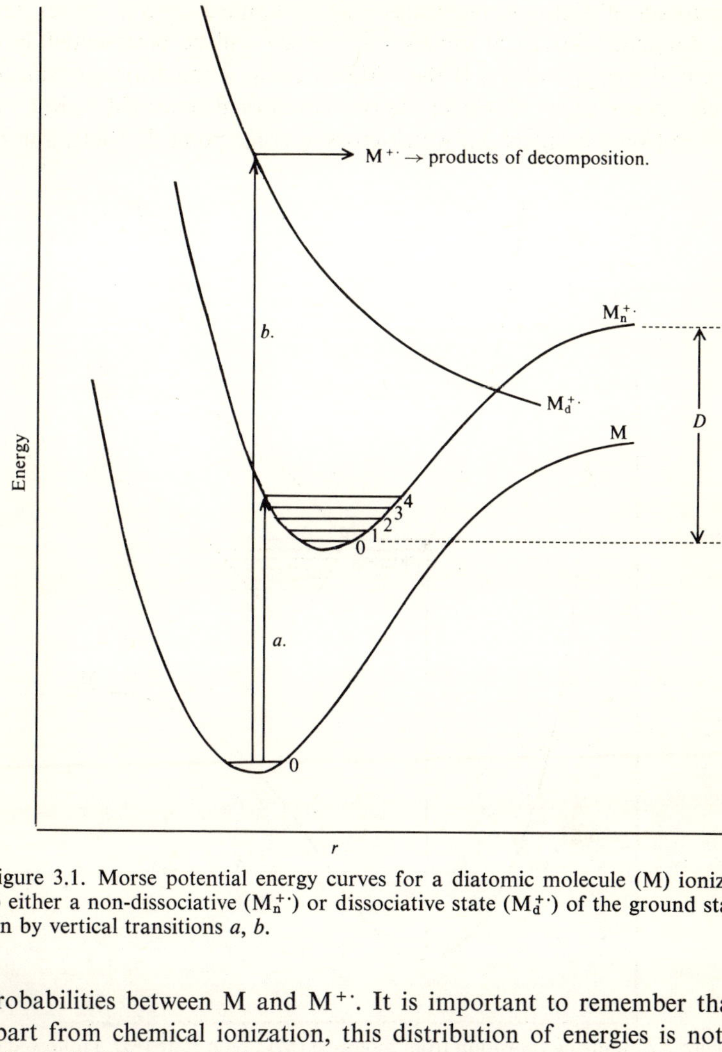

Figure 3.1. Morse potential energy curves for a diatomic molecule (M) ionized to either a non-dissociative ($M_n^{+\cdot}$) or dissociative state ($M_d^{+\cdot}$) of the ground state ion by vertical transitions *a*, *b*.

probabilities between M and $M^{+\cdot}$. It is important to remember that, apart from chemical ionization, this distribution of energies is not a 'Maxwell–Boltzmann' distribution since the latter applies to collisionally equilibrated species, and in mass spectrometry the ions are not so equilibrated because ion–ion and ion–molecule collisions are rare. Each ion behaves as a separate entity containing its own specific amount of energy which may be sufficient to induce fragmentation.

In complex molecules, the Morse curves must be replaced by potential energy surfaces, but similar principles to those used in the discussion of diatomic molecules may be applied. For a species having different vibrational modes, each mode will be represented by a potential energy surface. If the surfaces cross, a radiationless transfer of energy can occur. Consider the two vibrational states ($M_1^{+\cdot}$, $M_2^{+\cdot}$) of the ion ($M^{+\cdot}$) in figure 3.2 which cross at some point *X*. Ionization by

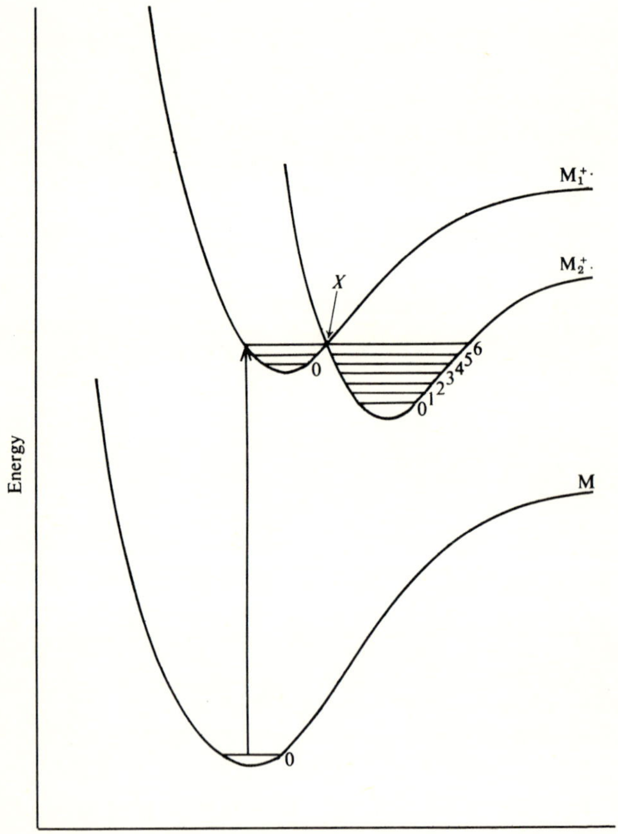

Figure 3.2. Ionizing transition from molecule (M) to ion state ($M_1^{+\cdot}$) followed by radiationless transition at *X* to ion state ($M_2^{+\cdot}$).

vertical transition from the $\nu = 0$ level of the molecule to the $\nu = 3$ level of the ion state ($M_1^{+\cdot}$) can occur, but the two states ($M_1^{+\cdot}$, $M_2^{+\cdot}$) coincide at *X* and crossing from state $M_1^{+\cdot}$ to state $M_2^{+\cdot}$ may take place. In state $M_2^{+\cdot}$, the internal energy transferred from $M_1^{+\cdot}$ may be above the dissociation energy so that the ion fragments. In complex molecules there are many such energy surfaces by which transfer of energy may occur and so, after ionization, equilibration of the excess of internal energy in the ion can take place rapidly by these surface crossing processes. There may be insufficient excess of energy to cause fragmentation and this results in a stable ion. Various complications to this basic approach to ionization and fragmentation must be considered.

The only ionization discussed so far corresponded to removal of the most loosely bound electron to give a vibrationally excited ion in its electronic ground state. At ionizing energies above the first ionization potential it is possible to remove electrons from lower orbitals to give electronically excited ions. Figure 3.3 shows the electron configuration of a ground state ion (removal of an electron from orbital ψ_1) and an excited state ion (removal of an electron from ψ_3). Figure 3.4 represents

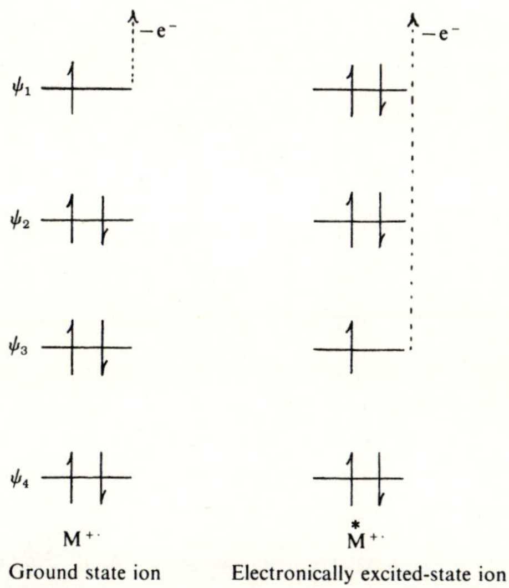

Figure 3.3. Orbital levels showing ionization to ground state molecular ion ($M^{+\cdot}$) and an electronically excited state ion ($\overset{*}{M}^{+\cdot}$).

the formation of these ions from the molecule by vertical ionizing transitions. In figure 3.4, the curve $M^{+\cdot}$ represents the ground state ion and $\overset{*}{M}^{+\cdot}$ the electronically excited state ion. The excited state ion may lose or redistribute its energy by two main processes: (i) internal electron redistribution with concomitant emission of radiation to give

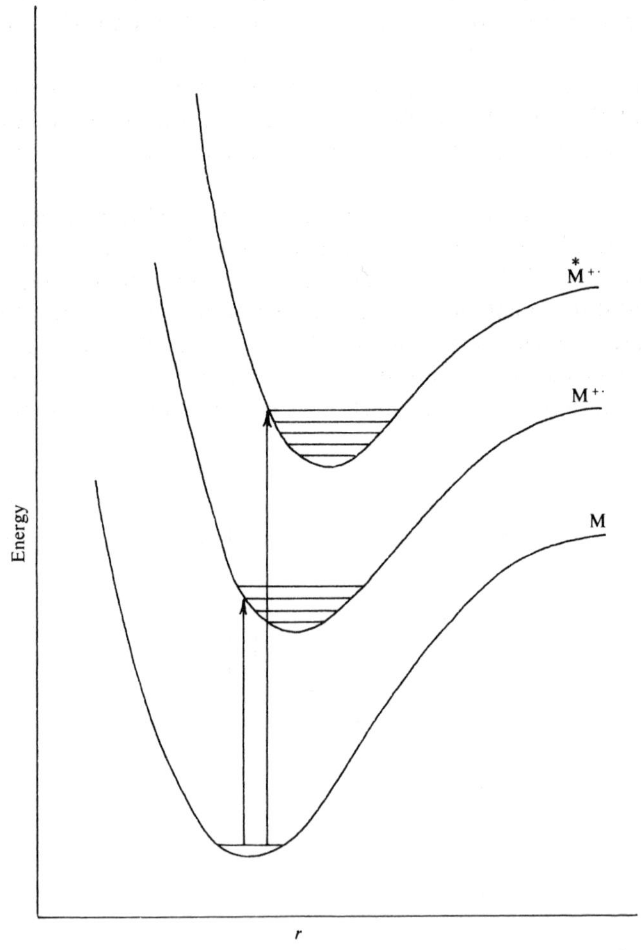

Figure 3.4. Ionizing transitions from the molecule (M) to the ground state ion ($M^{+\cdot}$) and to an excited state ion ($\overset{*}{M}^{+\cdot}$), corresponding to the electron configurations shown in figure 3.3.

a ground state ion and (ii) radiationless potential surface crossings as in figure 3.2 whereby the electronic energy in the excited state ion can be transformed into vibrational energy in the ground state ion. Thus, either by radiation of energy or by radiationless transfer of energy, the excited state ion can revert eventually to a ground state ion having an excess of vibrational energy. There are many possible excited states of ions which may be reached on ionization. Energy conversion by surface crossing does not necessarily lead simply to a ground state ion since if there is sufficient energy to cause decomposition from any excited state, then decomposition will occur from that state. Because the time for a bond vibration (*c.* 10^{-13} seconds) is short compared with the maximum lifetime of an excited state ion (*c.* 10^{-8} seconds) many bond vibrations with the possibility of radiationless state crossing are possible within this lifetime. The lifetime of an excited state ion (*c.* 10^{-8} seconds) is considerably shorter than the time (*c.* 10^{-5}–10^{-6} seconds) an ion spends in the ion chamber. There is therefore ample time for the excess of electronic energy in an ion to be converted into an excess of vibrational energy in a lower electronically excited state. As a general working hypothesis in mass spectrometry it is usually supposed that this less excited state from which most reaction occurs is the ground state of the ion. The process may be represented schematically as follows for molecules (M) interacting with ionizing radiation and fragmenting to yield ions found in the mass spectrum:

$$n.\mathrm{M} + \text{energy} \xrightarrow[\text{seconds}]{10^{-15}\text{–}10^{-16}} \underbrace{a.\mathrm{M}^{+\cdot} + b.\overset{*}{\mathrm{M}}{}_1^{+\cdot} + c.\overset{*}{\mathrm{M}}{}_2^{+\cdot} + \cdots}_{} + n.\mathrm{e}$$

$$\Big\downarrow \; < 10^{-7} \text{ seconds (internal energy conversion)}$$

$$\underbrace{a.\mathrm{M}^{+\cdot} + b.\mathrm{M}_1^{+\cdot} + c.\mathrm{M}_2^{+\cdot} + \cdots}_{}$$

$$\Big\downarrow \; 10^{-10}\text{–}10^{-5} \text{ seconds (fragmentation)}$$

$$\underbrace{p.\mathrm{M}^{+\cdot} + q.\mathrm{A}^{+} + r.\mathrm{B}^{+} + \cdots}_{}$$

$$\Big\downarrow \; \text{(ions accelerated from ion source)}$$

$$\text{mass spectrum}$$

$\overset{*}{\mathrm{M}}{}_1^{+\cdot}, \overset{*}{\mathrm{M}}{}_2^{+\cdot}, \ldots$ represent electronically excited molecular ions, $\mathrm{M}_1^{+\cdot}$, $\mathrm{M}_2^{+\cdot}, \ldots$ vibrationally excited molecular ions, and $\mathrm{A}^+, \mathrm{B}^+, \ldots$ fragment ions.

Removal of one electron from a molecule leaves an ion in the doublet state (one unpaired spin $\therefore 2S + 1 = 2$). During ionization by electron impact, only the total spin of the system (electron plus molecule) needs to be conserved so that an ion may also be formed in its quadruplet state (three unpaired spins). Just as triplet states have longer lifetimes than singlet states, it is likely that quadruplet states have longer lifetimes than doublet states. Hence, fragmentation may take place from relatively long-lived excited state ions.

Broadly a picture emerges of ionization producing molecular ions in different energy states, rapid internal energy conversion to produce ions with individual amounts of excess of energy, and fragmentation taking place at different rates depending on the electronic state of and excess of vibrational energy in each ion. The distribution of internal energies may be represented as in figure 3.5. Some ions have insufficient

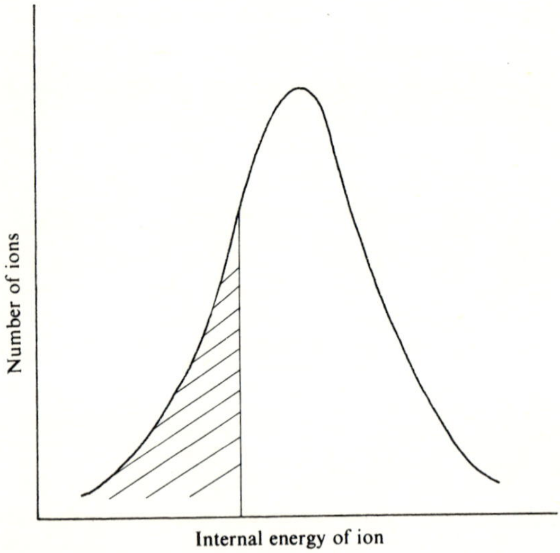

Figure 3.5. Distribution of internal energies in ions. The true shape of the curve would be governed by the cross-sections for all the various ionizing transitions. The hatched portion indicates the numbers of ions with insufficient energy to fragment.

energy to fragment (figure 3.5) and the proportion of them increases as the energy of the ionizing radiation decreases to the ionization potential. For this reason, reducing the electron beam energy in an electron-impact source reduces the amount of fragmentation observed in the mass spectrum. It is possible to consider a mean or average effect when discussing the total behaviour of many such ions, but this situation of isolated species reacting independently of one another is quite unlike the mean effect resulting from collisional processes in solution or normal gas phase chemistry. In solution chemistry, collision processes leave molecules with sufficient energy to react or decompose. In mass spectrometry, the ion initially produced does not normally collide with other species and usually decomposes endothermally if sufficient energy is transferred during ionization.

It is informative at this stage to consider the effects of the various methods of ionization on the final appearance of a mass spectrum. In field ionization, it is believed that only the most loosely bound electron is removed to give a ground state ion having little excess of vibrational energy. Further, the ions are collected very soon after ionization because of the high accelerating voltage used so that relatively few bond vibrations can occur. Because of these two factors, the likelihood of much fragmentation occurring is small and it is found experimentally that field-ionization yields mass spectra which normally have abundant molecular ions but relatively few fragment ions. Field-ionization spectra are therefore useful in mixture analysis, where molecular ions need to be easily distinguished, and for comparison with those electron-impact spectra which give weak molecular ions and much fragmentation. The field-ionization and electron-impact spectra of pent-1-ene are compared in figure 3.6.

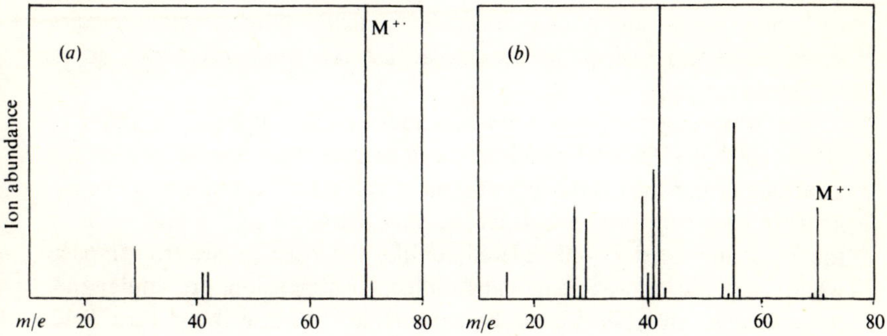

Figure 3.6. Mass spectrum of pent-1-ene determined by (*a*) field, and (*b*) 70-volt electron-impact ionization. Note the more prominent molecular ion and reduced abundances of fragment ions in spectrum (*a*) compared with spectrum (*b*).

Electrons accelerated through a potential of V volts have a de Broglie wavelength of $\sqrt{150/V}$ Å e.g. 75-volt electrons have a wavelength of $\sqrt{2}$ Å ($\sqrt{2}/10$ nm) corresponding to short wavelength radiation. During the approach of an electron to a molecule, the impacting electron waves and the electric field of the molecule mutually distort one another. The distorted electron wave can be considered to be composed of many different sine waves and some of these component waves will be of the correct frequency (energy) to interact with molecular electrons. Thus, the impacting electron may lead to electronic excitation in the molecule by promoting an electron from a lower to a higher orbital (this is the same effect as in ultraviolet spectroscopy). Similarly, a molecular electron may be promoted to an outer orbital (compare this with vacuum ultraviolet Rydberg bands) or, finally, an electron may be ejected from the molecule altogether to leave a positive ion. Other electronic effects in the molecule are also observed but there is insufficient space to deal with them here.

The term 'electron-impact' is misleading because an electron is so small in molecular terms that it would have difficulty 'hitting' any part of a molecule it met. It is better to think of the electron as passing close to or even through a molecule rather than of any 'impact' taking place. The effects of the impacting electron on the molecule are felt as it approaches and there is some distortion of the molecule during the transition to an ion i.e. the ionizing transitions are not strictly vertical and allow more states to be reached than by photo-ionization. Also, compared with photo-ionization, more spin states of the ion are accessible by electron-impact ionization. Therefore, in comparison with both field- and photo-ionization, ionization by electrons results in more excess energy in the molecular ion and fragmentation is enhanced.

The wavelength of the radiation required for photo-ionization is about 1000 Å (100 nm) which is much greater than normal molecular dimensions and means that a molecule is subjected to a uniform electric field. In quantum mechanical terms, only the effect of the radiation on the molecule need be considered, unlike the case of electron-impact when the radiation wavelength and molecular dimensions are similar and mutual quantum effects must be considered. Because the electric field of the photons is essentially uniform over the whole molecule, almost no molecular distortion occurs during ionization. The ionizing transitions are very fast (10^{-16} seconds) and are strictly vertical in accord with the Franck–Condon principle. Also, because of spin conservation from molecule to ion, fewer states are accessible than by electron-impact ionization. In comparison with the latter, much less vibrational energy appears in a molecular ion after photon impact and, like field-ionization spectra, photo-ionization spectra yield abundant molecular ions and relatively few fragment ions.

Chemical-ionization involves the collision of two bodies, an ion and a molecule, which are massive compared with an electron; the collisional interaction is much slower, allowing equilibration of energy between the two colliding species. Suppose the ionic species $A^{+\cdot}$ is formed by electron-impact ionization of A, then at the relatively high pressures used in the ion source the ions $A^{+\cdot}$ collide with neutral A molecules with equilibration of thermal energy (if the $A^{+\cdot}$ ion is atomic, as for example argon, then no equilibration of vibrational and rotational energy between $A^{+\cdot}$ and A is possible and only charge exchange can occur). If a new charged species, B^{+}, is formed by ion–molecule collision between $A^{+\cdot}$ and a different molecule B then these ions can also collide with other neutral A molecules leading to equilibration of vibrational and rotational energies. In chemical-ionization therefore it is believed that the ions possess fairly closely defined amounts of energy similar to those common in ordinary chemical reactions. Since the molecular ions formed by chemical-ionization may contain relatively small excesses of vibrational energy, reduced fragmentation and prominent molecular ions are often observed when compared with electron-impact spectra. The molecular ions resulting from chemical-ionization can be even-electron species derived by addition or subtraction of hydrogen during ionization or odd-electron species resulting from direct charge exchange (see §5.9).

$$MH + e \longrightarrow MH^{+\cdot} + 2e;$$ electron-impact produced odd-electron ions

$$MH + A^{+\cdot} \longrightarrow MH^{+\cdot} + A;$$ charge exchange produced odd-electron ions

$$MH + AH^{+} \begin{cases} \nearrow MH_2^{+} + A \\ \searrow M^{+} + AH_2 \end{cases}$$ hydrogen transfer produced even-electron ions

Figure 3.7 shows a comparison between the chemical-ionization and electron-impact ionization mass spectra of n-propyl propionate. The molecular ion is scarcely visible in the electron-impact spectrum but the chemical-ionization method gives an intense (M + H) ion peak.

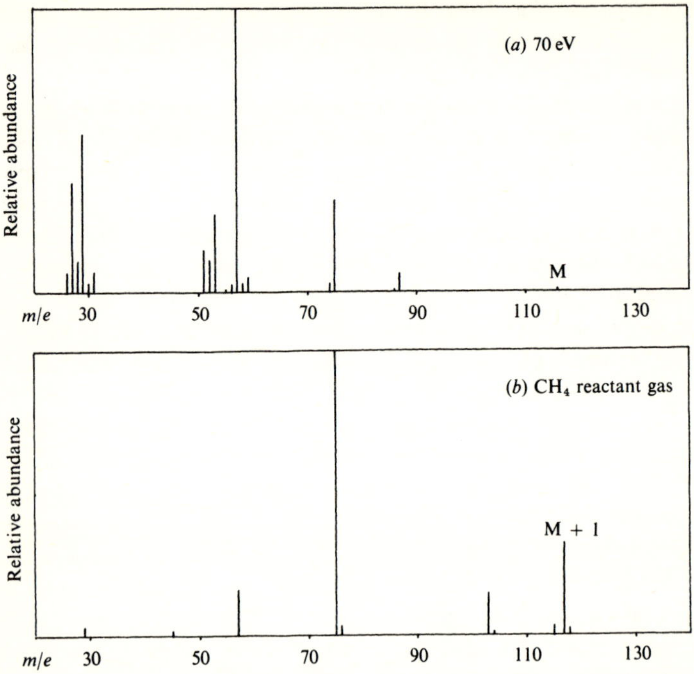

Figure 3.7. Comparison of electron-impact (*a*) and chemical-ionization (*b*) mass spectra of n-propyl propionate (mol. wt. 116).

In general, chemical-ionization may be used to advantage with sugars, peptides, nucleosides, alkaloids, terpenes, and steroids.

It is essential to remember that although the structure of a molecule put into a mass spectrometer may, at least in principle, be determined by other physical methods (e.g. X-ray analysis), after ionization it is not necessarily correct to assume the ion has the same structure as the original molecule. Immediately after the ionization step, the original molecule continues to vibrate as an ion and this vibration may lead rapidly to a structure quite different from that of the original molecule. There is ample experimental evidence to suggest that frequently re-arrangement does occur rapidly after ionization.

3.3. Quasi-equilibrium theory (**QET**). A quantitative treatment of mass spectrometric fragmentation has been attempted using the mathematical expression contained in equation (3.1). In this expression, the internal energy (E) in the ion is distributed amongst all the available oscillators (mean frequency, $\bar{\nu}$) and rotators (B) and the rate of decomposition (k_J) in any particular way may be calculated (Rosenstock

$$k_J = \frac{\bar{\nu}\cdot\left(\frac{\pi h\bar{\nu}}{B}\right)^{L/2} \sum_{P=1} \binom{N-L-1}{P}\left(\frac{1}{\sigma_P^{L/2}\Gamma(P+\frac{1}{2})}\right) \times \left(\frac{E-E_J}{h\bar{\nu}}\cdot\sigma_P - \frac{P-1}{2}\right)^{P+L/2}}{\sum_{Q=1}\binom{N}{Q}\left(\frac{\sigma_Q}{(Q-1)!}\right)\left(\frac{E}{h\bar{\nu}}\cdot\sigma_Q - \frac{Q-1}{2}\right)^{Q-1}} \qquad (3.1)$$

et al. 1952; Vestal, Wahrhaftig, and Johnston, 1962). A weighting factor or frequency coefficient (σ) is given to each vibrational level and L, M, N, P, Q are integers. For all but very simple molecules, the expression is too complex to be evaluated without the aid of a computer. The theory has been used with success to predict the mass spectra of some alkanes and a few other simple compounds. Some of the success may only be apparent since in applying the theory it is necessary to assume that the structure of a molecular ion is the same as that of the neutral molecule it came from. There is a simplified version of equation (3.1) which is useful for discussing mass spectra in a semi-quantitative fashion and can be used with complex molecules without the need of a computer. The simple expression giving the rate of decomposition (k) of an ion is shown in equation (3.2), where ν is a frequency factor, E the excess of internal energy in the ion, E_0 the energy of activation for the decomposition, and N the number of oscillators in the ion. The frequency factor is a kind of measure of the entropy factor in the decomposition; for a simple bond cleavage for example it is put equal

$$k = \nu\left[\frac{E-E_0}{E}\right]^{N-1} \qquad (3.2)$$

to the bond vibration frequency which may be obtained from an infrared spectrum. For a rearrangement reaction, a particular spatial disposition of atoms is probably required and it may be expected that ν would be at a lower frequency than for simple bond cleavage (Field and Franklin, 1957). The number of oscillators in a molecule containing n atoms is given by the expression $(3n-6)$; often it is found that not all the oscillators appear to be equally effective therefore leading to a calculated value for k which is too large, and the number of oscillators (N) is usually divided empirically by three for better results. The part of the expression (3.2) containing the energy terms, $(E-E_0)/E$, is a dimensionless quantity and energy can be considered in any convenient units e.g. electronvolts.

The simplified rate equation (3.2) has been used to discuss the fragmentation of organic molecules in a general sense. As simple examples of its use the following may be mentioned: the greater the energy transferred during ionization i.e. $(E - E_0)/E$ increases, the faster an ion decomposes; for two alternative fragmentations of an ion having similar activation energies the rates are proportional to the frequency factors i.e. $k_1/k_2 = \nu_1/\nu_2$; if the frequency factors are similar the rates of fragmentation are approximately proportional to the activation energies; comparison of the rate for a simple bond fission reaction with the rate for fission accompanying rearrangement suggests that the rearrangement process may become more prominent as the internal energy of the ion decreases (§4.5). It should be noted that these rates (k) refer to individual ions with internal energy (E) and that the rates are not equivalent to the rates found in solution chemistry which are the result of an equilibration process amongst millions of molecules. It is however possible in a qualitative sense to consider the behaviour of an ion having an 'average' excess of energy and extrapolate its behaviour to all the other ions of the same structure but slightly different internal energies. In this way the simple QET equation (3.2) can be used to illustrate some general principles and it is used in this way in the later discussions of ion rearrangement processes (§4.5) and metastable ion transitions (§5.1). The approximate equation (3.2) is very inaccurate at low internal energies (E) and should not be used in these regions.

In using the simplified QET equation for a process such as $A^+ \rightarrow B^+$, subsequent fragmentation of B^+ is often ignored. Particularly at higher internal energies, these further fragmentations are important and the abundance of ion B^+ (or A^+) cannot be simply related to its rate of formation (or decomposition). As stressed in §1.1, the abundance of an ion in a mass spectrum depends upon both its rate of formation and its rate of decomposition, and care should be exercised in the use of the QET equation if observed ion abundances are to be related to calculated rates of formation or decomposition. Abundances of ions considered theoretically in this way are unlikely to reflect very closely the observed abundances unless some account is taken also of the distribution of excess energy in the ions. The actual distribution of excess energy may be such that the use of an 'average' excess, as suggested above, would lead to gross differences between theory and practice.

3.4. Qualitative theories. Discussion of mass spectrometric fragmentation for structural analysis must be carried out in terms of the structure of the intact molecule before ionization because the structures of ions are unknown. Fragmentation of a molecular ion is described with minimum structural change at each step. For example, the loss of CH_3O from methyl benzoate can be described as simple bond cleavage, without knowledge of the actual structures of the ions, simply by considering a minimum of structural change during ionization and decomposition necessary to remove this fragment.

$$[C_6H_5CO{\cdot}OCH_3]^{+\cdot} \longrightarrow C_6H_5CO^+ + CH_3O^{\cdot}$$

Conversely, loss of CH_3O from a substance of unknown structure may be taken to imply that such a unit is present without any knowledge of ion structures other than their elemental compositions. Thus, for structural analysis of organic compounds by mass spectrometry it is only necessary to investigate and collate the mass spectra of many compounds of known structure and to use this knowledge in the analysis in an empirical fashion. This empirical approach is common in organic chemistry and has proved extremely useful in ultraviolet, infrared, and nuclear magnetic resonance spectroscopy. Because the actual structures of ions are unknown it is not possible to discuss adequately the mechanisms of mass spectrometric fragmentations, and still less the intervention of intermediates and transition states. Nevertheless, 'mechanisms' for mass spectrometric fragmentation are frequently published and sometimes 'supported' by evidence of isotope labelling experiments. Whether or no such mechanisms are ever proved correct or incorrect has no effect on the present value of the empirical approach to mass spectrometry for structural analysis. Two qualitative theories of mass spectrometry which have proved of considerable value as aids to structure determination are described. The main attribute of these theories is that they classify the many types of mass spectrometric fragmentation into a few categories of behaviour. Knowing these categories, it is easier to discuss the mass spectrometric fragmentation of new compounds of known or unknown structure.

Stabilities of fragmentation products. For the fragmentation of an ion to take place it must possess an excess of internal energy sufficient to exceed the activation energy for the reaction considered (E_f in figure 3.8). It has been demonstrated for many fragmentation reactions of ions, and is believed to be generally approximately correct, that the activation energy for the reverse reaction (E_r) is almost zero or at

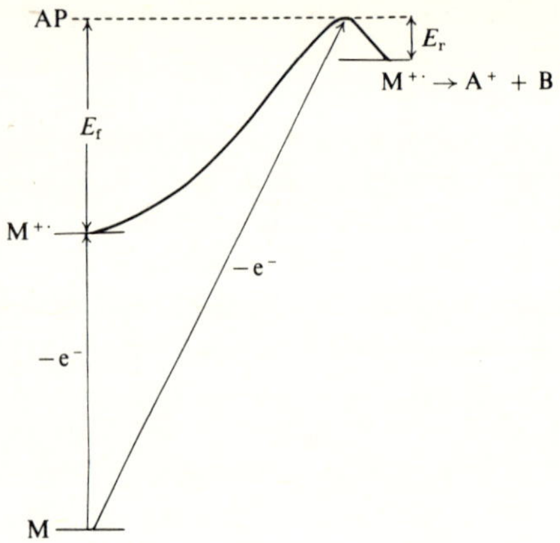

Figure 3.8. Energy diagram showing the relationship between the activation energies (E_f, E_r) for the forward and backward reactions in the decomposition of the ion ($M^{+\cdot}$) to fragments (A^+, $B^\cdot$). The appearance potential for the process is marked (AP).

least very small. By Hammond's postulate (1955) for reaction kinetics, it may be supposed, for such endothermic fragmentations, that the transition state will resemble the products of reaction and therefore that the stabilities of the products will, to an approximation, determine the course of the fragmentation. Consider two alternative decompositions (3.3*a*, *b*) of the molecular ion of methyl benzoate. Figure 3.9

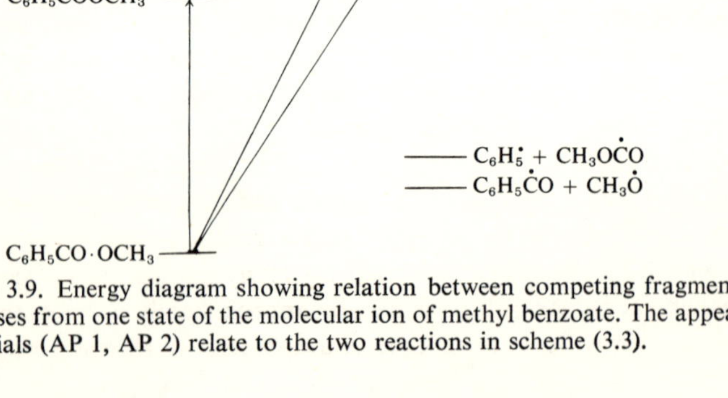

Figure 3.9. Energy diagram showing relation between competing fragmentation processes from one state of the molecular ion of methyl benzoate. The appearance potentials (AP 1, AP 2) relate to the two reactions in scheme (3.3).

$$[C_6H_5CO \cdot OCH_3]^{+\cdot} \nearrow C_6H_5CO^+ + CH_3O^\cdot \quad (3.3a)$$

$$\searrow C_6H_5^+ + CH_3O \cdot CO^\cdot \quad (3.3b)$$

shows the relative stabilities of these products with respect to the molecular ion. If the transition state does resemble the products ($E_r \approx 0$) and if the frequency factors (equation 3.2) are similar then reaction (3.3*a*) will be favoured over (3.3*b*) as found experimentally. For the proper use of this method it is necessary to take account of the heats of formation of all the products of fragmentation. When the ejected neutral particle is a stable molecular species like CO or CO_2, its large negative heat of formation should not be ignored, since it will lead to enhanced fragmentation by reducing the energy required for the process compared with similar processes in which the neutral fragment ejected does not have a large negative heat of formation. Quite often only the stabilities of the ionic products are considered. For example, the fragmentation of an amine, $RCH_2NR'_2$, beta to nitrogen is often ascribed to 'resonance stabilization' of the positive charge in the fragment ion $CH_2NR'_2$; the 'stability' of the neutral part R is ignored. The 'wiggly' line is drawn to indicate which bond is considered to

$$[R \wr CH_2NR'_2]^{+\cdot} \longrightarrow R^\cdot + \overset{+}{C}H_2NR'_2 \longleftrightarrow CH_2{=}\overset{+}{N}R'_2$$

$$(\text{For } E_r = 0, \text{ then } E_f = \Delta H_f(R^\cdot) + \Delta H_f(\overset{+}{C}H_2NR'_2) - \Delta H_f(RCH_2\overset{+\cdot}{N}R'_2))$$

break in the process. A charged species is drawn as a structure inside square brackets. An even-electron ion is marked by a (+) sign outside the brackets and an odd-electron species (radical–cation) by a ($+^\cdot$) sign. Many types of mass spectrometric fragmentation have been classified by Biemann (1962) into a few general schemes, some of which are outlined in table 3.1. These classifications are empirical, imply actual knowledge of ion structures, and consider fragmentation in terms of the structure of the original molecule. Two examples will suffice to illustrate the argument:

(i) n-Alkane molecular ions decompose extensively to give a series of fragment ions (C_nH_{2n+1}) separated by 14 mass units:

$$[C_nH_{2n+2}]^{+\cdot} \longrightarrow CH_3^+, C_2H_5^+, C_3H_7^+, \ldots$$

However, a branched alkane shows enhanced bond fission adjacent

TABLE 3.1 *Examples of empirically classified fragmentation modes (Biemann, 1962)*†

Type A_3: 

Type A_5: 

Type B: 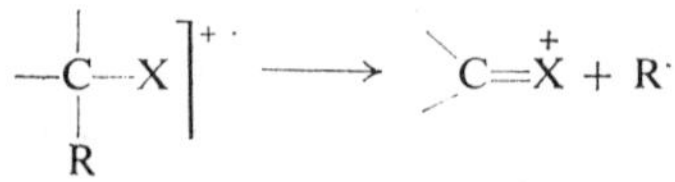

† These and other reaction types are classified slightly differently in this book; the Biemann classification represents the first, and still valid, comprehensive effort in this way of approach to reaction paths in mass spectrometry. Type A_3 represents 'allylic' cleavage, Type A_5 and B represent cleavage adjacent to the α-carbon atom (see §4.4).

to the branching point and this has been ascribed to the increased stability of secondary or tertiary cations compared to the primary ones generated by n-alkanes:

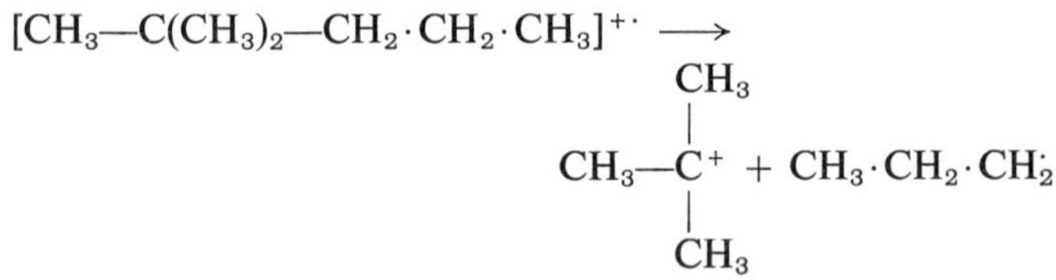

(ii) Bond cleavage of the cinnamyl compound (1) takes place predominantly at the position shown and this is supposed to be due to the stability of the resulting cinnamyl cation (2). The subsequent loss of H_2 from this cation may also be ascribed to cyclization with formation of the stabilizing indenyl cation (3):

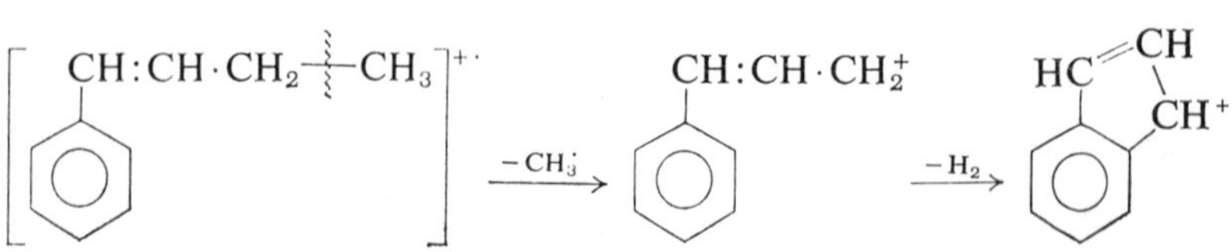

(1) (2) (3)

Charge localization. An alternative theory of mass spectrometric fragmentation applied extensively by Djerassi (see for example, Budzikiewicz, Djerassi, and Williams, 1967) supposes that, after ionization, the charge on the molecular ion may be considered localized at some particular place in the ion. For an aliphatic amine $RCH_2NR'_2$, this site of charge localization is the nitrogen atom because one of the 'lone-pair' electrons on nitrogen would be easiest to remove. The charged site is then considered to 'trigger' fragmentation by one- or two-electron shifts. A double-headed arrow indicates a two-electron

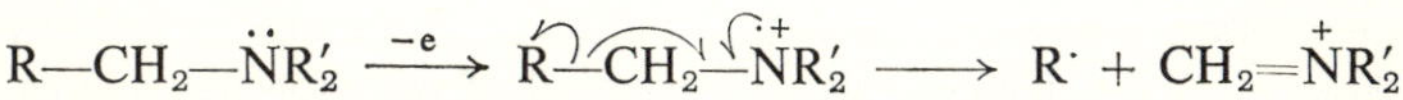

shift and a single-headed arrow, or 'fish-hook', a one-electron shift. Charge stabilization by the product ion is therefore a 'bonus' making the reaction more favourable. It has also been argued that it is specifically the presence of the radical site which is the driving force for the reaction (McLafferty, 1966). To apply the charge localization treatment it is only necessary to decide where the charge is localized (this may not in fact be at all obvious in a complex molecule). It is unnecessary to classify fragmentations by type, although this is often done, and therefore the method provides both a valuable *aide-mémoire* for collecting large amounts of information on mass spectrometric fragmentation and a means of assessing the possible direction of this fragmentation in compounds for which mass spectra have not been determined. Similarly, using these same principles for the interpretation of the mass spectra of compounds of unknown structure affords a convenient guide when considering alternative possible structural features. The losses of an ethyl radical from the epoxide (1) and of C_2H_2O from acetanilide (2) are used to illustrate the method:

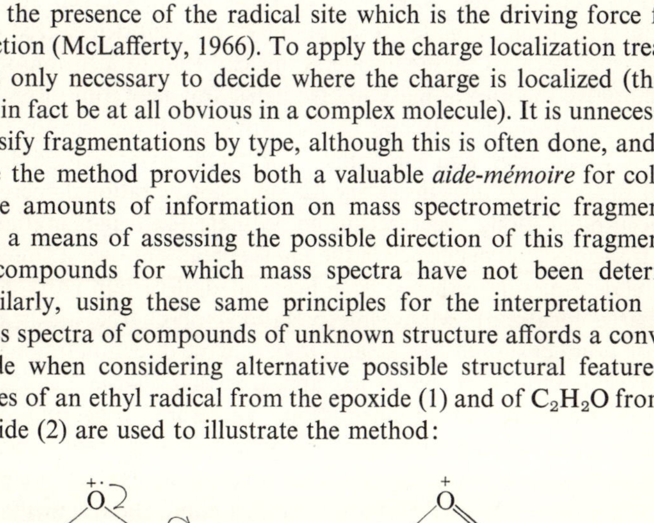

(1)

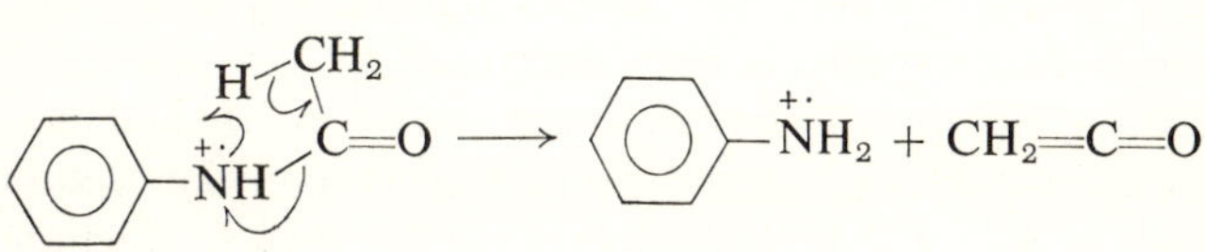

(2)

Comments on the two qualitative theories. For a theory to be worthy of the name, the underlying concept should have a truly predictive capability. This criterion does not apply to either of the qualitative theories since neither is predictive but both rest heavily on argument by analogy or comparison with previous experience. From a theoretical point of view, the first hypothesis based on Hammond's postulate of the transition state is possibly more satisfying although less fashionable. The second theory requiring a charge localized site to trigger fragmentation would appear to be opposite to Hammond's postulate. There is no reason to suppose that triggering is necessary for fragmentation along any particular pathway. Electrons in a molecule are held in molecular orbitals and removal of one electron affects the whole structure to various degrees. Even 'lone-pair' electrons are not strictly localized on atoms although this is a good approximation for ground state properties of organic molecules (Dewar and Worley 1969). For example, the 'lone-pair' electrons on the oxygen atom in methanol have been estimated to be about 25 per cent delocalized over the molecule. Perhaps these considerations can lead to a fusion of the two qualitative theories of mass spectrometric fragmentation. Removal of an electron from a molecular orbital will change the various bond vibration frequencies and strengths to different extents in different parts of the ion. It is only necessary for sufficient vibrational energy to be concentrated along the reaction co-ordinate for fragmentation to occur. If the electron deficiency in the ion is spread out over a 'molecular' orbital it may nevertheless be concentrated at certain parts in the ion (charge localized). Many interpretations of charge localization really use product stability arguments to decide the direction fragmentation will take and, from this, it has appeared that charge localization leads to 'triggering' of fragmentation. Therefore, triggering often appears as a deduction arising from charge localization theory when in fact, product stability has already been invoked to explain preferred modes of cleavage. By way of illustrating this point, the fragmentation of compounds C_2H_5X may be considered. Charge localization alone cannot determine whether α- or β-cleavage in C_2H_5X compounds (3.4) is more favourable but product stability arguments can. In fact, although bond strengths have been affected in different ways on ionization, Hammond's postulate may still be applied and the stabilities of products will largely determine the direction of fragmentation in the absence of a large activation energy for the reverse reaction and not forgetting the effects of frequency factors. It may be suggested therefore

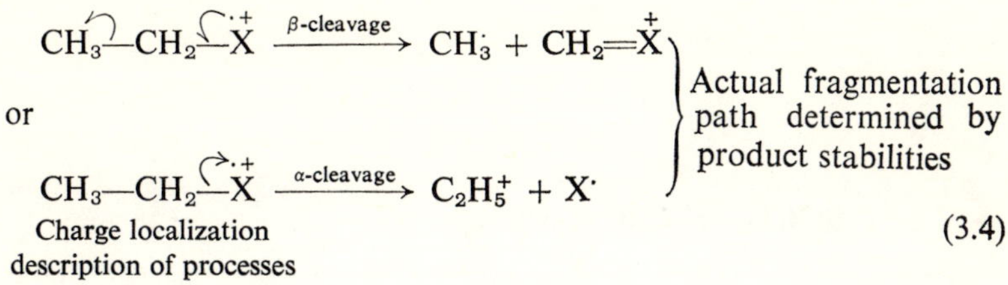

(3.4)

that charge localization can indicate which bonds in a molecule are most likely to be affected on ionization and that product stability arguments will then mostly determine the actual direction of fragmentation. From this point of view, the two theories are complementary and their conventions of the 'wiggly line' and 'fish-hooks' or 'arrows' will accordingly be used somewhat indiscriminately throughout this book as a means of 'electron book-keeping'.

4 Structure elucidation

4.1. Classifications of mass spectra. The mass spectra of many thousands of compounds of known structure have now been determined and published mostly in normalized tabular form. The ways in which such published spectra may be used for structure elucidation of 'unknown' compounds are described briefly.

Compilations of mass spectra. A large fraction of the thousands of recorded spectra is scattered throughout the literature in reports on research work. Quite often, the mass spectra are only incidental to the work reported and it can be a tedious task looking through the literature for particular ones. In the past there have been attempts by various agencies and sets of research workers to collate mass spectra in published volumes or on magnetic tape. More recently, a start has been made to collect and make available from one central organization all published mass spectra (Mass Spectrometry Data Centre – see bibliography).

If the mass spectrum of a substance of unknown structure has been obtained it is obviously useful to be able to sort through compilations of published spectra and compare them with it. In this comparative way it may be possible to identify the 'unknown' without further analysis of the mass spectrum. The task of searching through large compilations of mass spectra and comparing each one with the one under investigation is a prodigious task when done by hand. Fortunately, this sort of effort is almost trivial for a computer which can provide an answer in a short time. Of course, the classification of mass spectra in this way is useless if the mass spectrum of the material to be identified has not been determined previously. On the other hand, the method is extremely useful where a relatively limited range of compounds is examined repetitively. Provided most, if not all, the mass spectra in the range required are available then positive identification of an unknown may quickly be made simply by comparison with the library of spectra.

Functional groups. Classifications of mass spectrometric fragmentations based on the functional group approach so successful in organic chemistry have been equally successful in mass spectrometry. Thus, the fragmentation reactions of aliphatic ketones for example, may be classified together and any fragmentations which appear to be common features of the presence of a ketone group are noted particularly. If an unknown substance is suspected of being a ketone, then its mass spectrum can be examined for features of fragmentation pathways reported for other ketones, and the classifications by functional group make this task much easier. Frequently, the effects of changes in molecular structure on the fragmentation behaviour of the functional group are included in the classifications and provide very useful extra information. For example, a comparison of the effects on fragmentation of a ketone group in ring and straight-chain compounds can help to decide the environment of a ketone group in an unknown substance.

The functional group approach has its limitations. There may not be sufficient difference between the effects of functional groups to allow a positive identification to be made. Changes in structure can produce large changes in the 'standard' effects of functional groups. The molecule may contain more than one functional group and if these interact in the fragmentation process then the effect could be to mask the 'standard' effects of each of the groups. These problems are not exclusive to mass spectrometry but occur frequently in chemistry and illustrate the inherent dangers of relying exclusively on one particular technique for providing an answer to a problem.

There is neither the scope in, nor is it the intention of, this book to attempt more than this summary of the functional group approach to the interpretation of mass spectra. In general, the scheme is very close to similar ones long familiar to organic chemists in infrared, ultraviolet, and nuclear magnetic resonance spectroscopy, and in the classification of reactions of organic compounds. Accurate mass measurement of the molecular ion in a mass spectrum will give the molecular formula of the compound and this in itself often affords a good indication of the type of functional group present. Several excellent books have been published which classify and discuss mass spectra in detail on the basis of functional groups and the reader is recommended to read them for further information (Beynon, Saunders and Williams, 1968; Budzikiewicz, Djerassi and Williams, 1964*a*,*b*, 1967; Hill, 1966; McLafferty, 1966; Reed, 1966).

Characteristic m/e values. It is often observed that a certain feature of molecular structure will give rise to a characteristic peak in the mass spectrum such that the presence of an ion at that *m/e* value may be taken as good evidence for the feature. This approach is not the same as that based on functional group since the latter concentrates on fragmentation pathways whereas one based on characteristic *m/e* values relies on the identification of one or two particular peaks. Thus, an ion at *m/e* 43 is often evidence for the presence of the grouping CH_3CO in a molecule; it is of course even better to determine the elemental composition of the ion at *m/e* 43 by accurate mass measurement to support the diagnosis. The presence of an ion at *m/e* 91 (C_7H_7) is almost classical in mass spectrometry and usually indicates the presence of either a benzylic residue in the molecule or a structure which readily rearranges to give the C_7H_7 ion. Similarly, an ion at *m/e* 30 (CH_4N) strongly indicates an amine, particularly a primary one.

Lists of such characteristic peaks have been published and they may be used in much the same way as for example nuclear magnetic resonance correlation charts (Beynon *et al.* 1968; McLafferty, 1963). Just as peaks near τ 2–3 in a proton magnetic resonance spectrum would suggest the presence of an aromatic structure, so too would the presence of an ion at *m/e* 91 in the mass spectrum. However, in general, these compilations of characteristic peaks do not have the same usefulness as the correlation charts used in ultraviolet, infrared, and nuclear magnetic resonance spectroscopy because, in mass spectrometry, rearrangement reactions accompanying fragmentation are common, i.e. the original molecule is considerably disturbed on ionization. For example, although an ion of composition C_2H_3O at *m/e* 43 suggests the CH_3CO grouping in a molecule, it is found that many oxygen containing compounds without this grouping also give intense ions at this *m/e* value. As a mass spectroscopist gains experience from interpreting mass spectra so the usefulness of characteristic peaks increases.

The method is particularly valuable when closely similar sets of organic compounds are examined. For example, an ion at *m/e* 105 (C_7H_5O) is prominent in the mass spectra of molecules containing the grouping C_6H_5CO and on substitution of the benzene ring the corresponding ion appears at a different *m/e* value. Thus, benzophenone yields an intense ion at *m/e* 105 but 3-chloro-3′-methylbenzophenone gives ions at *m/e* 119 (C_8H_7O, or $CH_3 + C_7H_4CO$) and *m/e* 139, 141 (C_7H_4OCl, or $Cl + C_7H_4CO$):

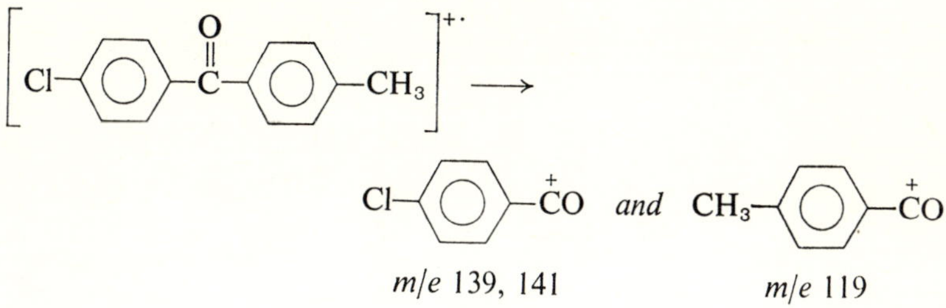

In the examination of the mass spectra of closely defined areas of natural product chemistry, characteristic peaks have played a leading role in structure determination. The examination of mass spectra of peptides, benzylisoquinoline alkaloids, indole alkaloids, and sterols by workers experienced in the spectra of these compounds will quickly yield much valuable information. Therefore, where closely related compounds are being examined it is worthwhile searching their mass spectra for peaks which are characteristic of particular common groupings, so that these groupings may be recognized in the mass spectra of similar compounds of unknown structure. For example, benzylisoquinoline alkaloids (1) give a prominent, characteristic ion (2) in their mass spectra which is due to ejection of the benzylic residue; trimethylcoclaurine ($R^1 = R^2 = R^3 = CH_3$) yields the ion (2) at *m/e* 206.

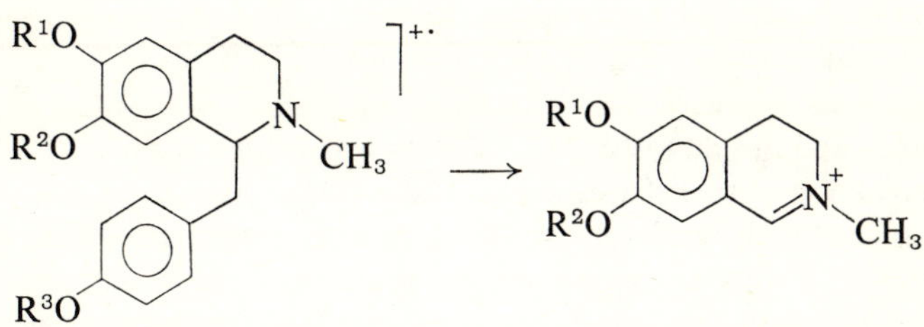

(1) (2)

4.2. Examination of the mass spectrum. The first step in examining a mass spectrum is usually to decide which peak represents the molecular ion. The spectrum is counted to increasing *m/e* values until no more ions are observed and this is then the molecular ion region. In assigning the molecular ion it is worth recalling that it is accompanied by isotope ions one or more mass units greater (M + 1, M + 2, etc.). For chlorine,

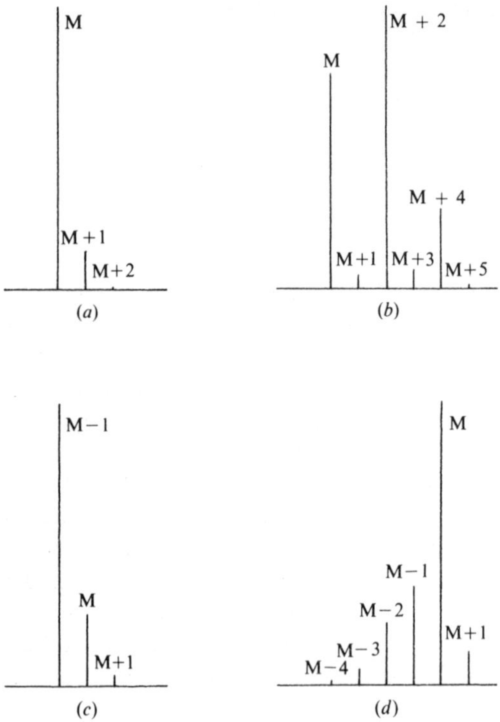

Figure 4.1. Comparison of the molecular ion ($M^{+\cdot}$) regions of (*a*) phenylacetylene, (*b*) 1-bromo-2-chlorobenzene, (*c*) 2-styrylpyridine, (*d*) biphenyl.

bromine, and sulphur the M + 2 ions are intense compared to the molecular ion (M). Figures 4.1*a*,*b* show respectively examples of the molecular ion region for a hydrocarbon with a normal molecular ion region, and a dihalo-compound with peaks up to M + 5. Very often, M − 1, and sometimes M − 2, M − 3, etc. ions are observed owing to losses of hydrogen from the molecular ion. Figure 4.1*c* illustrates the molecular ion region of 2-styrylpyridine which looks very similar to that of the hydrocarbon in 4.1*a*. However, the most intense ion in the nitrogen compound is the M − 1 peak and not the molecular ion. Sometimes these losses of hydrogen are so marked as to lead to the wrong assignment of the molecular ion because the M − 1 ion is so much more intense than the molecular ion. If such losses of hydrogen are suspected it is worthwhile examining the spectrum at lower ionizing energies and for metastable ions (see below, p. 53). Figure 4.1*d* shows the molecular ion region in biphenyl with the marked loss of four successive hydrogen atoms from the molecular ion.

Difficulties are also encountered where the molecular ion is 'weak' or even non-existent so that it becomes impossible to say exactly what is the molecular weight of the compound. Sometimes the difficulty can be resolved by examining the fragment ions and deducing the molecular weight. It is more satisfying to re-examine the mass spectrum after ionization with low electron beam energies, photons, or electric fields, or after chemical ionization since these methods often afford good molecular ions when under normal electron-impact conditions, only uncertain ones are found.

If only the molecular weight or molecular formula of the compound is required, then deciding on the molecular ion will be the only examination of the spectrum required. Even so, this can still lead to much valuable information. As has been shown above, the very appearance of the molecular ion region can furnish information. It is useful to remember that an *odd* molecular weight indicates an *odd* number of nitrogen atoms in the compound. Intense M + 2 ions would suggest the presence of chlorine or bromine, and less intense ones the presence of sulphur. An approximate idea of the number of carbon atoms in the molecule can be gained from the relative heights, h and $h + 1$ of the M and M + 1 ions respectively. Since ^{13}C is present in about 1.1 per cent natural abundance in organic compounds, the approximate number of carbon atoms is given by $100(h + 1)/(1.1h)$. For example, in the mass spectrum of naphthalene having ten carbon atoms, $(h + 1)/h = 10.9/100$ so that $(100 \times 10.9)/(1.1 \times 100) = 10$ to the nearest integer. The method is not very accurate for more than ten or twelve carbon atoms and may yield highly erroneous results if the compound contains nitrogen (contribution of ^{15}N to M + 1 peak) or if there are overlapping peaks due to impurities.

To obtain the molecular formula of a compound, it is necessary to measure the accurate mass of the molecular ion. Knowing the accurate mass, the molecular formula can be obtained from tables or by calculation (Beynon and Williams, 1963; Lederberg, 1964). Table 4.1 shows some of the elemental compositions possible at m/e 100, and if we suppose that the measured accurate mass of the molecular ion of an unknown compound was 100.0635, it can be seen from the table that this mass corresponds very closely to that calculated for $C_4H_8N_2O$. The error between calculated and observed masses is usually reported in parts per million. In the example cited, the difference between

TABLE 4.1 *Some elemental compositions at mass 100*

Composition	Accurate mass
$C_3H_6N_3O$	100.0511
$C_4H_8N_2O$	100.0637
$C_5H_{10}NO$	100.0762
$C_6H_{14}N$	100.1126
C_7H_{16}	100.1251
$C_2H_4OSi_2$	99.9700

calculated and observed values is 0.002 mass units i.e. the error is two parts in 1 000 637 which is about two parts per million. Errors of up to ±10 p.p.m. are usually acceptable but the smaller the error, the more certain the molecular formula. It may be mentioned here that accurate mass measurement is not a substitute for elemental analysis although the two complement each other very well. The mass spectrometer yields the exact molecular formula but may give little indication of the purity of a compound. The elemental analysis provides a cross-check on the purity of a compound and on its molecular formula.

Having obtained a molecular formula, it may then be used to calculate the number of double bond equivalents in the compound. For the commoner elements met in organic chemistry, this may be done very simply by substituting CH_3 for each halogen, CH_2 for each oxygen or sulphur atom, and CH for each nitrogen. For example, carrying out these substitutions on the molecular formula C_6H_6ONCl gives C_9H_{12}; a straight-chain alkane with nine carbon atoms would have a molecular formula C_9H_{20}. The difference between these formulae, $C_9H_{20} - C_9H_{12}$, is eight hydrogen atoms and therefore there are four double bond equivalents in C_6H_6ONCl.

During the initial examination of a mass spectrum its general appearance is frequently a guide to the nature of the compound under investigation. A spectrum exhibiting many fragment ions increasing in abundance towards low *m/e* values will suggest a predominantly aliphatic structure whereas a spectrum with few fragment ions, a good molecular ion, and doubly charged ions will suggest an aromatic structure. Figure 4.2 illustrates this point by comparing the mass spectra of naphthalene and valeraldehyde.

The next step in examining the mass spectrum is to note the major fragment ions and to attempt to elucidate the main fragmentation

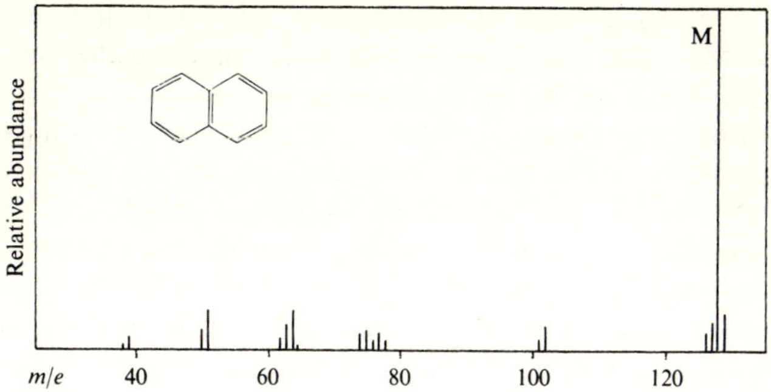

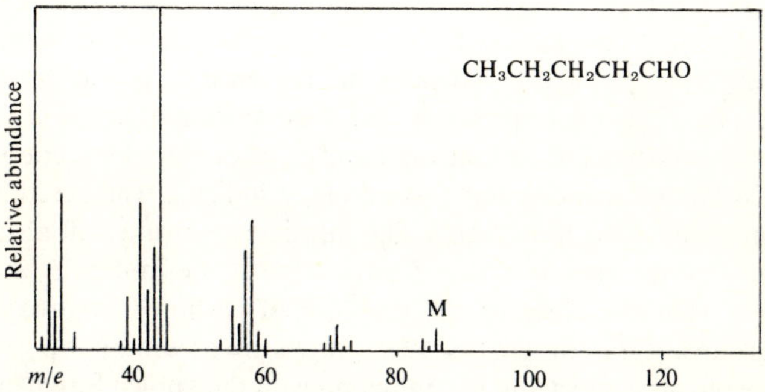

Figure 4.2. Comparison of mass spectra of an aromatic compound (naphthalene) and an aliphatic compound (valeraldehyde).

pathways. The major fragment ions may occur at 'characteristic' m/e values (see above) and give immediate information. If the fragment ions occur at odd m/e values they are even-electron species (§4.4) resulting from simple bond fission, whereas fragment ions at even m/e values are odd-electron species produced by multiple bond cleavage and suggest that rearrangement may have occurred. In this latter case extra care must be taken in interpreting the spectrum because some of the fragment ions may well arise from rearranged species.

It is best to check supposed fragmentation pathways by looking for metastable ions in the routine spectrum; recall that the position of the metastable ion (m*) is given by the expression m_2^2/m_1 where m_1 is the mass of the precursor ion and m_2 the mass of the fragment ion. The mass spectrum of ethylbenzene [M] (m/e = 106) shows strong peaks at [M − 15] (m/e 91) and [M − 41] (m/e 65). The presence of metastable ions at *m/e* 78.1 and *m/e* 46.4 proves that one fragmentation path at least is [M] → [M − 15] → [M − 41] since the calculated values for the metastable ions for these processes are *m/e* 78.1 ($91^2/106$) and *m/e* 46.4 ($65^2/91$). Often, metastable ions for suspected fragmentation reactions are not visible in the routine spectrum and it is then necessary to look for them by the 'defocussing' technique (see §5.1.).

Having obtained the molecular weight and the main features of the fragmentation of the compound under investigation, it may be possible to make tentative or even definite structural assignments. Comparison of the mass spectrum with those held in a library may allow immediate identification of the compound (§4.1). At this stage, it is advisable to assemble any other available physical data on the compound. Interpretation of a proton magnetic resonance spectrum will benefit from a knowledge of the molecular formula which gives the exact number of protons present and also indicates the presence of particular hetero-atoms. The ultraviolet spectrum will yield information on the likely degree of unsaturation so that the number of double-bond equivalents obtained from the molecular formula may indicate whether there are any rings, and if so how many. The infrared spectrum will also often give information on the state of saturation of the molecule and may give very valuable clues to the presence of carbonyl, hydroxyl, and other groups. If it is known that such groups are present, it is possible to examine the mass spectrum in conjuction with the known fragmentation behaviour of these functional groups (§4.1). If the infrared spectrum indicates the presence of a hydroxyl group, then reference to the fragmentation reactions of alcohols in general may reveal similar fragmentations in the spectrum of the unknown. It is very much a case of pooling all available information, using it logically, and calling on experience.

It may not be possible to proceed any further in the interpretation without knowledge of the elemental compositions of the fragment ions. Loss of 27 mass units from the molecular ion of C_6H_6ONCl may be HCN or C_2H_3 and only accurate mass measurement on the ion at [M − 27] will determine which it is. A simple mass spectrum containing few fragment ions requires only a little extra work to obtain the accurate masses of these ions and hence their elemental compositions. On the other hand, a complex mass spectrum requires many accurate mass measurements and it should be remembered that any peak at an integer *m/e* value may include ions of different compositions. At high resolution therefore some peaks will appear as doublets or triplets. A long list of elemental compositions corresponding to the major ions in a mass spectrum is difficult to sift for information of help in structural work. The problem has been recognized especially following the application of computers to accurate mass measurement (§5.2) since this technique generates a great deal of information which must be examined in detail. The commonest method of assembling all the data into some order is through the use of an *element map*. An example of one common type of these maps is given in figure 4.3 and relates to the mass spectrum

m/e CH	*m/e* CHO	*m/e* CHO_2	*m/e* CHO_3	*m/e* CHO_4	*m/e* CHO_5
		60 $C_2H_4O_2$			
		61 $C_2H_5O_2$			
	59 C_3H_7O	71 $C_3H_3O_2$			
	70 C_4H_6O	83 $C_4H_3O_2$ (d)	103 $C_4H_7O_3$ (d)		
69 C_5H_9	81 C_5H_5O	100 $C_5H_8O_2$			
71 C_5H_{11}	82 C_5H_6O				
	83 C_5H_7O (d)				
77 C_6H_5	94 C_6H_6O	109 $C_6H_5O_2$ (d)	128 $C_6H_8O_3$		
78 C_6H_6	95 C_6H_7O		129 $C_6H_9O_3$		
79 C_6H_7	96 C_6H_8O				
80 C_6H_8					
90 C_7H_6	105 C_7H_5O	124 $C_7H_8O_2$	139 $C_7H_7O_3$		
91 C_7H_7	106 C_7H_6O				
92 C_7H_8	107 C_7H_7O				
93 C_7H_9	108 C_7H_8O				
	109 C_7H_9O				
102 C_8H_6	119 C_8H_7O	135 $C_8H_7O_2$			
103 C_8H_7	120 C_8H_8O	137 $C_8H_9O_2$			
	121 C_8H_8O				
	130 C_9H_6O	147 $C_9H_7O_2$	165 $C_9H_9O_3$		
	131 C_9H_7O	148 $C_9H_8O_2$	166 $C_9H_{10}O_3$		
		149 $C_9H_9O_2$	167 $C_9H_{11}O_3$		
		151 $C_9H_{11}O_2$			
		152 $C_9H_{12}O_2$			
		158 $C_{10}H_6O_2$	176 $C_{10}H_8O_3$	194 $C_{10}H_{10}O_4$	
		159 $C_{10}H_7O_2$		195 $C_{10}H_{11}O_4$	
		161 $C_{10}H_9O_2$	179 $C_{10}H_{11}O_3$		
		162 $C_{10}H_{10}O_2$	180 $C_{10}H_{12}O_3$		
		163 $C_{10}H_{11}O_2$			
			190 $C_{11}H_{10}O_3$	208 $C_{11}H_{12}O_4$	226 $C_{11}H_{14}O_5$

Figure 4.3. Element map of genipin. The ions are arranged in order of increasing complexity from left to right and top to bottom.

of a natural product, genipin. The ions are arranged in an array of increasing complexity of elemental compositions from left to right, and increasing *m/e* values from top to bottom. Thus, the molecular ion is found on the right of the map; doublet and triplet peaks are marked (d, t). The actual intensities of the peaks are not given in figure 4.3, but this information is frequently included also on element maps.

Other methods of element mapping have been proposed. Inspection of the map in figure 4.3 shows for example that the ion at *m/e* 129 cannot arise from *m/e* 147 (difference of 18 mass units which could have been due to H_2O) because the former has one more oxygen atom than the latter. The minimum number of carbon atoms associated with all five oxygen atoms is eleven; with three oxygen atoms it is four (*m/e* 103). The column headed CHO in the element map shows that nine carbon atoms are associated with one oxygen (*m/e* 131). The formula for genipin (4.1) shows which atoms these must be. Thus, if the structure of genipin had been unknown significant structural features could be gleaned from the map.

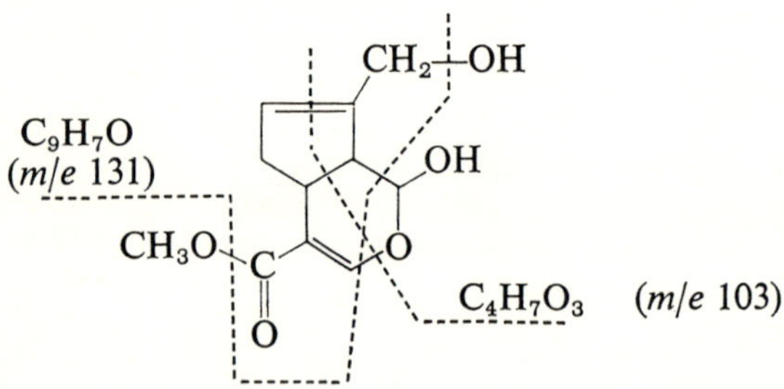

(4.1)

These maps provide a convenient means of presenting all the information relating to the elemental compositions of the ions in a spectrum but usually do not give data for metastable ions. With the advent of semi-automatic recording of metastable ions in the first and second field-free regions (§5.1), the use of *metastable ion maps* in conjunction with element maps is likely to increase.

4.3. Modification of mass spectra. After initially examining a mass spectrum it may be desirable to modify it by changing the ionizing energy or ion source temperature. A complex spectrum can be simplified by using lower energy radiation for ionization. A change in the appearance of a spectrum with time or the presence of unexplained peaks may be due to thermolysis of the sample, or to the sample being a mixture.

Ionizing energy. Field- and photo-ionization sources usually operate at a single ionizing energy as with the 21.21 eV helium line, but electron-impact sources have continuously variable electron-beam energies. Electrons from a hot filament are accelerated through a potential of

V volts and, in mass spectrometers used for organic chemistry, the voltage (V) ranges from about 5 to 100 volts. By convention, *standard* mass spectra are obtained with 70-volt electrons because at this energy the yield of ions is near maximum and almost constant over a small voltage range. A typical graph showing ion yield as a function of electron energy is given in figure 4.4. A maximum yield of ions is

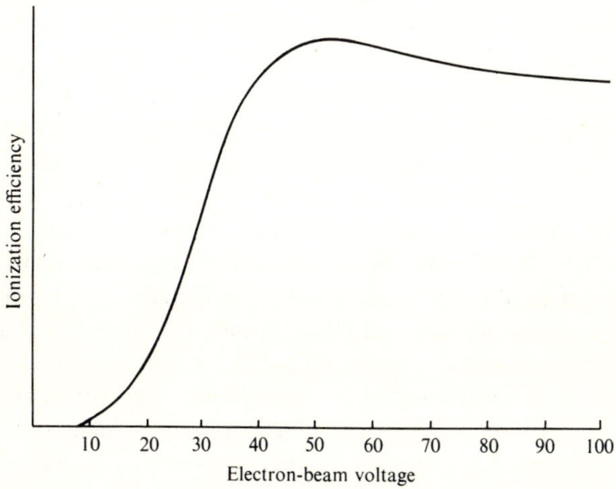

Figure 4.4. Typical curve illustrating variation in ion yield from a compound with increasing electron-beam voltage.

often obtained at electron energies of 50 volts rather than 70 but the former energy is not used for standard work because near the maximum, a small variation in voltage may lead to a pronounced change in the mass spectrum. There is little significance attached to the conventional figure of 70 volts for the ionizing electrons and a mass spectrum can be obtained at any energy above the ionization potential of the compound examined. For greatest sensitivity it is better to measure spectra at 50–70 volts.

The ionization potentials of most organic compounds lie between 7 and 11 electronvolts and from the ionization potential up to an electron-beam energy of about 20 volts, not only does the yield of ions increase but also the *appearance* of the mass spectrum changes appreciably. At, and just above the ionization potential, only molecular ions are formed providing the molecular ion is not in a dissociative state. As the electron-beam energy is increased more vibrational energy is produced in the molecular ion by internal energy conversion processes outlined earlier (§3.2.). The molecular ion then fragments and fragmentation becomes increasingly complex as the initial internal energy of the molecular ion increases with increase in electron-beam energy. The increasing electron energy means that higher and higher excited states of an ion ($\overset{*}{M}{}_1^{+\cdot}$ for example in figure 4.5) can be reached by vertical transitions from the molecule M. At some point, vertical transitions to highly excited states of the ion ($\overset{*}{M}{}_2^{+\cdot}$ for example) are not possible and no more energy can be transferred from the impacting electron. This point is not reached abruptly and maximum transfer of energy from the impacting electron is reached relatively slowly. Thus, as the energy of the impacting electron is increased, the energy which is imparted to a molecule on ionization increases rapidly at first and then much more slowly after about 20 eV. At the same time, the probability of ionization continues to increase so that the actual *yield* of ions continues to increase from about 20 to about 50 electronvolts although the maximum transfer of energy occurs at around 20 electronvolts. Therefore, although more and more ions are formed up to 50–70 electronvolts, little extra fragmentation is observed after 20 electronvolts. Experimental evidence suggests in fact that only about 5–6 electronvolts of energy above the ionization potential of a compound are transferred by 70-volt electrons. The effect on the appearance of the mass spectrum is as follows. At the ionization potential, molecular ions are formed; up to about 20 eV the mass spectrum becomes increasingly complex as more and more energy is transferred during ionization; after 20 eV and up to 50–70 eV the appearance of the mass spectrum changes little but the total yield of ions increases leading to greater sensitivity in the mass spectrometer. Figure 4.6 illustrates changes in the mass spectrum of ethyl thiolbenzoate with increasing electron beam energy.

Conversely, a mass spectrum can be simplified considerably by measuring it at electron-beam energies below 20 volts and simplification of a complex mass spectrum in this way may lead to a better understanding of the fragmentation paths. Figure 4.7 shows the effect on the mass spectrum of genipin of reducing the electron-beam energy from 70 to about 10 volts. The spectrum is greatly simplified and two ions at *m/e* 78, 96 of particular importance for its interpretation 'emerge' from the clustered fragment ions at low *m/e* values (typical of aliphatic compounds) in the higher energy spectrum.

The simplifying effect of reducing the electron-beam energy may

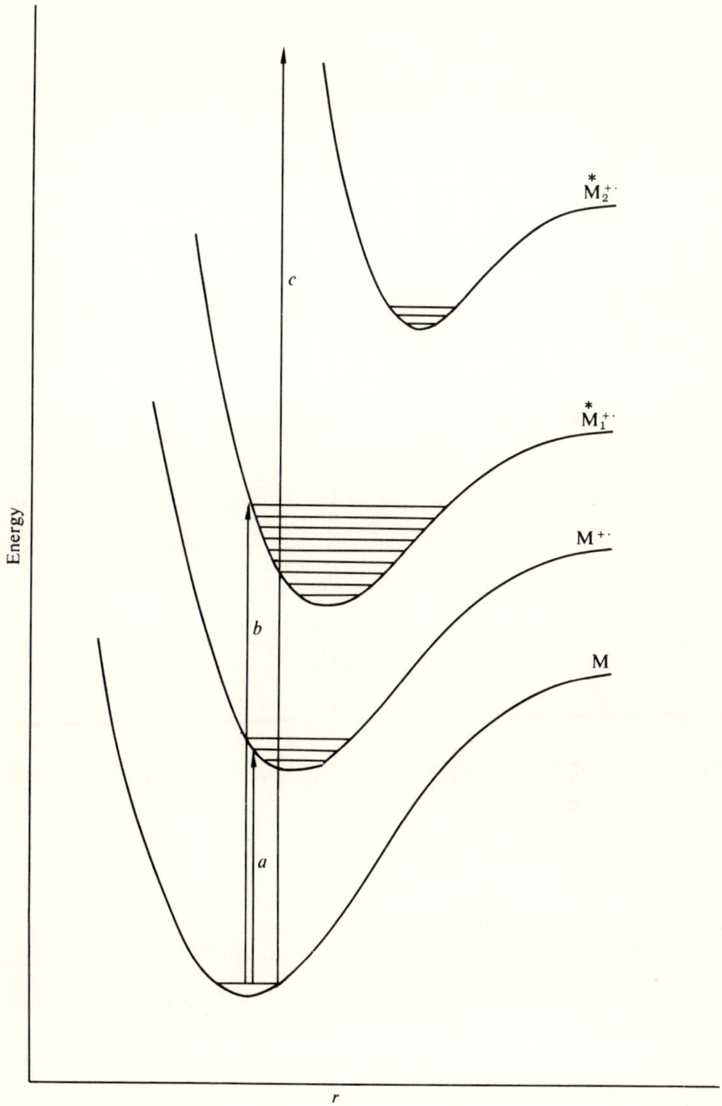

Figure 4.5. Morse curves showing ionizing transitions to excited states $\overset{*}{M}{}_1^{+\cdot}$, $\overset{*}{M}{}_2^{+\cdot}$ for a molecule, M.

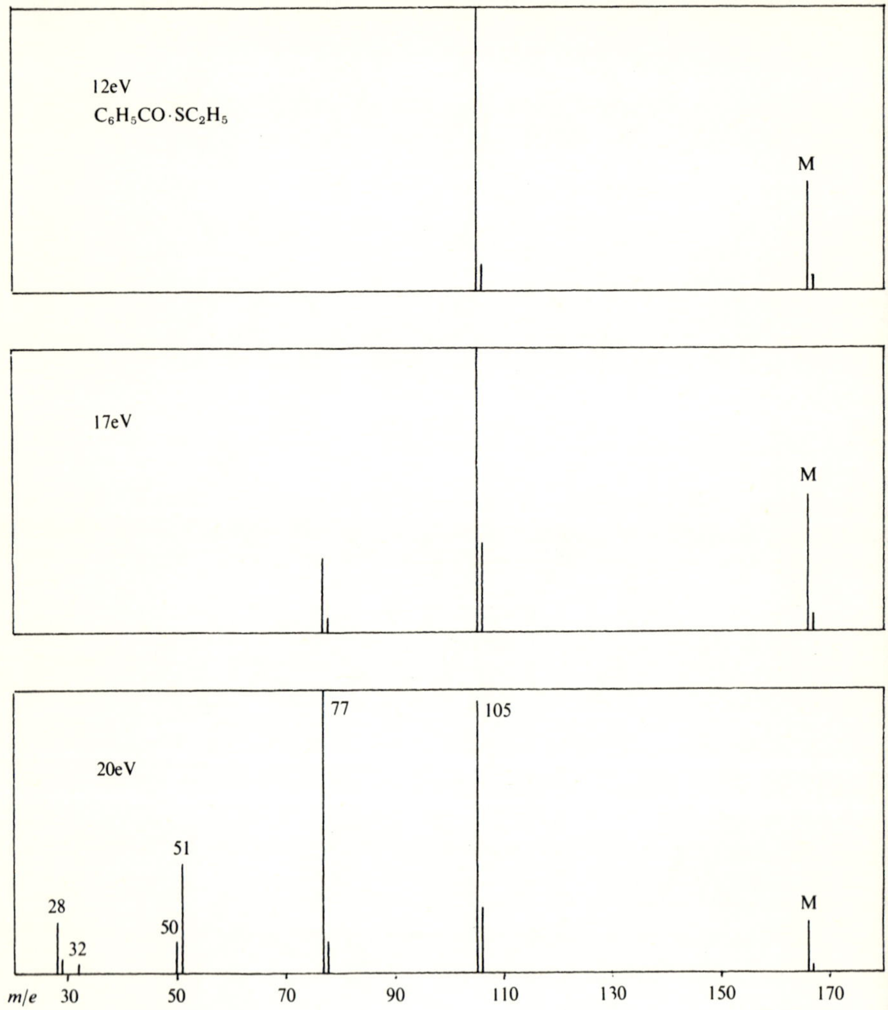

Figure 4.6. Changes in the mass spectrum of ethyl thiolbenzoate with increasing electron-beam energy.

be used in conjunction with reduction of the ion source temperature (see below) to afford even simpler mass spectra.

Temperature. The thermal energy imparted to a sample by high temperatures in the inlet system or ion source may have profound effects on the mass spectrum. The effects observed may result from

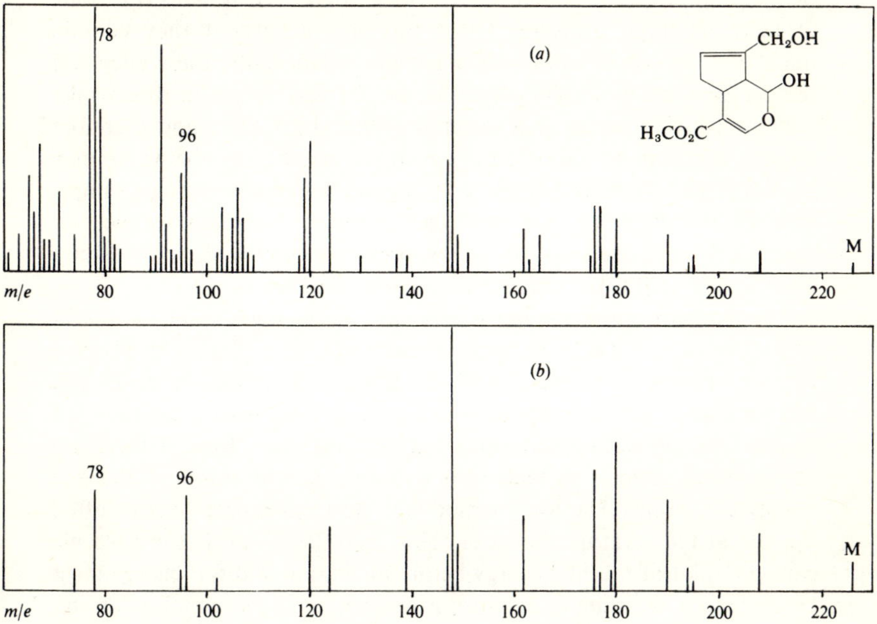

Figure 4.7. Mass spectra of genipin at (*a*) 70, (*b*) 10 eV. The lower energy spectrum is much simpler and peaks at *m/e* 78, 96 remain prominent.

actual changes in molecular structure or may be caused by the excess of vibrational and rotational energy gained by the molecule before it is ionized.

It is often necessary to heat a sample to achieve a great enough vapour pressure in the ion source for mass spectrometry. On heating, a compound may react uni- or bimolecularly leading to new compounds. The mass spectrum of the sulphonylhydrazide (1; 4.2.) was complicated by the presence of the azine (2; 4.2) formed from it thermally. Effects

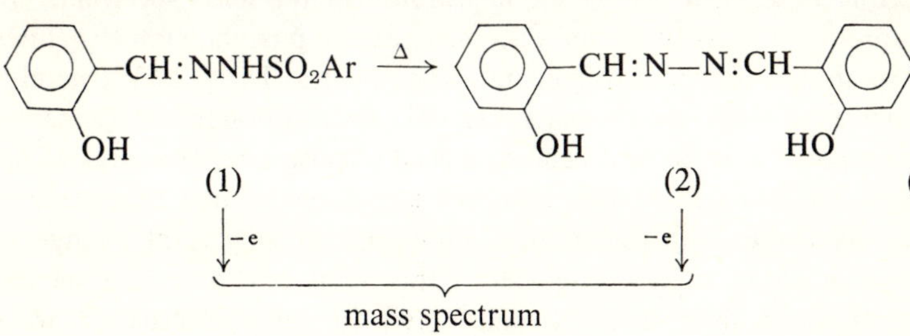

(4.2)

of this kind are more often found with heated inlet systems operating at relatively high pressures (10^{-2} torr) particularly if they contain metallic parts which can catalyze reactions. Normally, these effects of temperature on molecular structure are far less severe in direct inlet systems which operate at a lower pressure (10^{-6} torr) and therefore require less heat for vaporization. A direct insertion probe which may be independently heated or cooled is a good device for getting a sample into the ion chamber in a controlled manner. Although bimolecular reaction is much less likely at the low pressures employed, unimolecular reaction can still take place inside the ion chamber.

The temperature of the ion source can affect a mass spectrum markedly other than by actual chemical reaction taking place. On entering the ion chamber, molecules are not necessarily ionized immediately, but because of the low vapour pressure (long mean-free path) they can collide with its walls many times before ionization. Even if there are no chemical effects, at each wall collision, equilibration of thermal energy can occur. The ion chambers of electron-impact sources often operate at temperatures between 150°C and 250°C so that a molecule can gain a lot of thermal energy before ionization and this energy must be added to that imparted during ionization i.e. the molecular ion contains excesses of energy from the hot source and from electron-impact. In practice, it is found that the mass spectra of aromatic compounds having fairly rigid structures show little evidence of this thermal effect but aliphatic structures frequently show greatly enhanced fragmentation because of it. Figure 4.8 compares the partial mass spectrum of a peptide at two temperatures of the ion source and it can be seen that the intensity of the molecular ion peak is greatly reduced at the higher temperature. Electron-impact ion sources are available which can be cooled externally thereby giving some control over the temperature to which a sample is subjected.

4.4. Effects of structure on the appearance of the mass spectrum. In interpreting a mass spectrum it is convenient to postulate ion structures to explain the observed fragmentation pathways. The actual structures of ions in a mass spectrometer are unknown although the elemental compositions may be, and hence the need to postulate structures. If the compound investigated has a known structure the mass spectrum is often interpreted by assuming a minimum of structural change at each fragmentation step. To illustrate this point, part of the fragmentation of acetanilide may be considered in three alternative ways

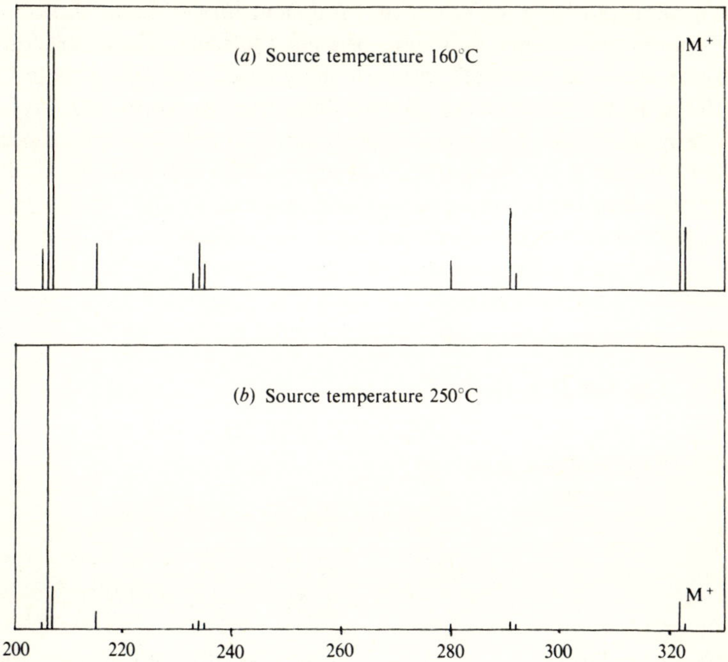

Figure 4.8. Partial mass spectrum of the peptide Z.Val.Gly.OMe at (*a*) 160°C (*b*) 250°C. Both spectra are normalized to *m*/*e* 206.

(4.3*a*,*b*,*c*). The first way (*a*) simply shows ionization and fragmentation in terms of *m*/*e* values and gives little information. The second (*b*) gives the same pathway in terms of elemental compositions of the ions; this affords much more information since the compositions of the ions and of the unit lost at the fragmentation step are known. The third way (*c*) adds postulate to experimental fact by assuming or postulating

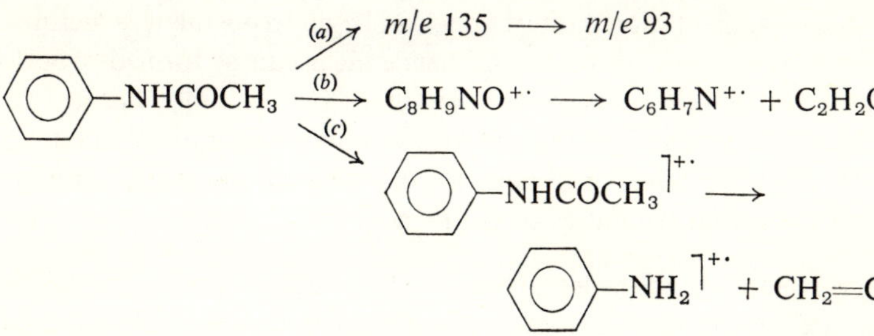

(4.3)

the structures of the molecular and fragment ions to be as shown i.e. with minimum change from the original molecule. These structures may or may not eventually prove to be correct, but this is quite immaterial to the convenience of describing fragmentation pathways in this way. An added advantage to postulation of ion structures appears when comparing the mass spectra of groups of compounds for similar fragmentation behaviour since this becomes much more obvious from structure relationships rather than elemental composition relationships. It is easier to see the similarity in behaviour between diethylmethylamine (1) and *n*-amylamine (2) from the postulated ion structures (4.4*a*) than from the corresponding elemental compositions (4.4*b*).

$$\left.\begin{array}{l} \underset{(1)}{CH_3CH_2N(CH_3)CH_2CH_3]^{+\cdot}} \longrightarrow \\ \qquad CH_3^{\cdot} + CH_2{=}\overset{+}{N}(CH_3)CH_2CH_3 \\ \underset{(2)}{CH_3CH_2CH_2CH_2CH_2NH_2]^{+\cdot}} \longrightarrow \\ \qquad CH_3CH_2CH_2CH_2^{\cdot} + CH_2{=}\overset{+}{N}H_2 \end{array}\right\}(a) \qquad (4.4)$$

$$\left.\begin{array}{l} C_5H_{13}N^{+\cdot} \longrightarrow CH_3^{\cdot} + C_4H_{10}N^{+} \\ C_5H_{13}N^{+\cdot} \longrightarrow C_5H_{11}^{\cdot} + CH_4N^{+} \end{array}\right\}(b)$$

The postulation of ion structure does cause difficulty in discussing thermochemical aspects of mass spectrometry. If it is necessary to postulate or assume an ion structure then that postulate or assumption is built into any thermochemical result achieved and therefore introduces uncertainty. Discussion of fragmentation mechanism is also clouded by those assumptions. It is one thing to postulate ion structures as an aid to describing fragmentation but it is quite another thing altogether to discuss detailed aspects of fragmentation mechanisms in terms of assumed structures!

The usefulness of postulating ion structures to describe fragmentation pathways is also applicable in a reverse sense for structural work. Knowing a fragmentation pathway one can begin to postulate structures which might behave in this way and hence ideas can be formed regarding the structure of an unknown compound. For example, knowing a molecular ion ejects $C_2H_3O_2$ (59 mass units) one can postulate a CO_2CH_3 (methyl ester) grouping in the original molecule without knowing the actual structures of the ions.

4.5. Fragmentation of hydrocarbons. The carbon–carbon bonds in alkanes are weaker than carbon–hydrogen bonds. In an alkane molecular ion, assuming it to have a similar structure to that of the molecule before ionization, the strengths of C—C and C—H bonds are greatly reduced. Table 4.2 compares the C—C and C—H bond strengths in

TABLE 4.2 *Bond strengths in ethane and the ethane radical–cation in kcal/mole and, in parentheses, kJ/mole*

	D_{C-C}	D_{C-H}
C_2H_6	83 (347)	96 (402)
$C_2H_6^{+\cdot}$	44 (184)	26 (109)

ethane and the ethane ion. If all the C—C bonds in the molecular ion have similar strengths it might be expected that alkane ions would cleave at them almost indiscriminately. Figure 4.9 shows the mass

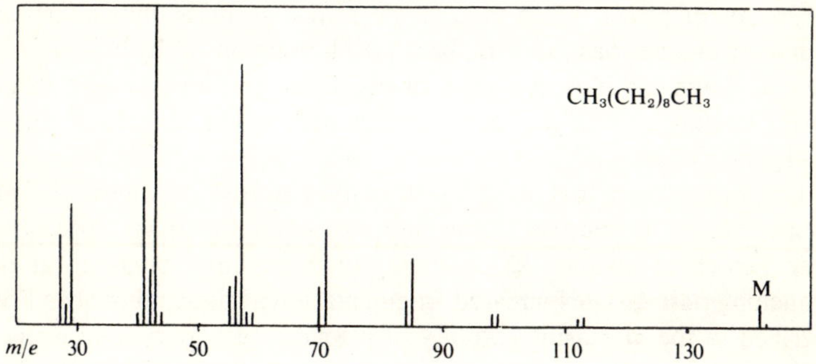

Figure 4.9. Typical electron-impact mass spectrum of an alkane, n-decane.

spectrum of n-decane in which can be seen bunches of ions approximately 14 mass units apart spreading from the molecular ion. The differences of 14 mass units are not due to elimination of units of CH_2 (an energetically unfavourable process) but rather to fragmentation at different places in the molecular and fragment ions (scheme 4.5). For an alkane, fragment ions appear at *m/e* values corresponding to the compositions C_nH_{2n+1}. This simple view of the mass spectra of alkanes is complicated by rearrangement processes and excess of internal energy in the ions. The excess of energy causes rapid fission of the C—C bonds in all ions resulting in a clustering of ions of small mass near

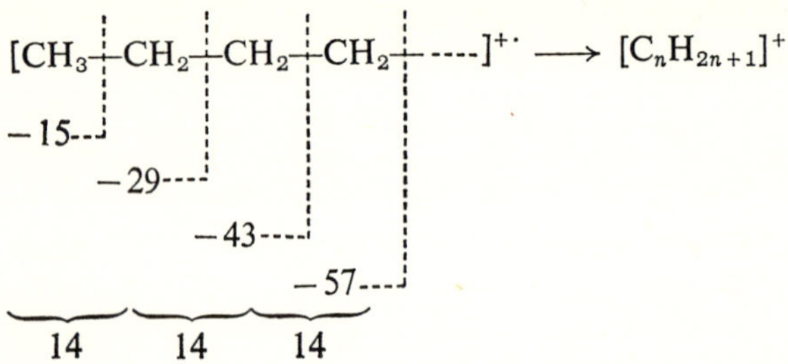

(4.5)

m/e 29, 43, 57 and a rapid falling off in the intensities of the peaks at higher *m/e* values up to the molecular ion which is generally of very low abundance. The excess of energy imparted to the molecular ion may be reduced by measuring the mass spectrum at low electron-beam energies and by using a cooled ion source. Under these conditions the mass spectrum is much more uniform with all fragment ions of similar intensities and without the bunching at low *m/e* values.

Groups of peaks rather than single peaks, 14 mass units apart, are found in alkane mass spectra because elimination of hydrogen from various fragment ions can take place. Thus, the ions at *m/e* values corresponding to C_nH_{2n+1} are accompanied by less abundant ions at C_nH_{2n} and C_nH_{2n-1}.

The supposition that product stability governs the mode of fragmentation (§3.4) receives some support from the mass spectra of branched-chain alkanes which show enhanced bond breaking at the branching points. The increased fragmentation at these points has been ascribed to the increasing stability of primary, secondary, and tertiary carbonium ions. Figure 4.10 shows the mass spectrum of 3,3-dimethylhexane compared with that of its isomer, *n*-octane. There are greatly increased abundances of ions at *m/e* 71, 85 corresponding to bond scission adjacent to the tertiary carbon centre in the branched-chain isomer. The phenomenon is useful for determining branching points in alkanes.

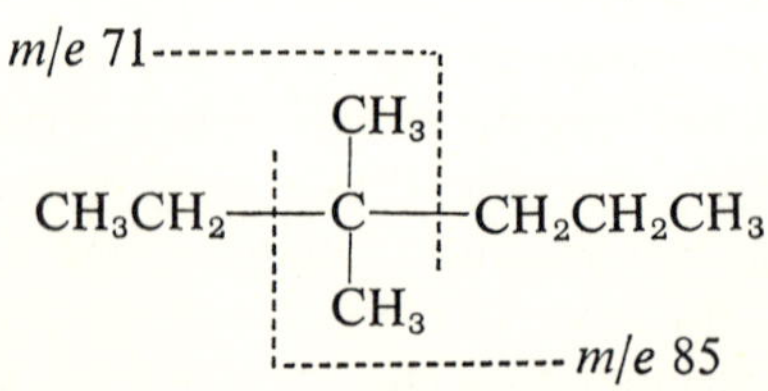

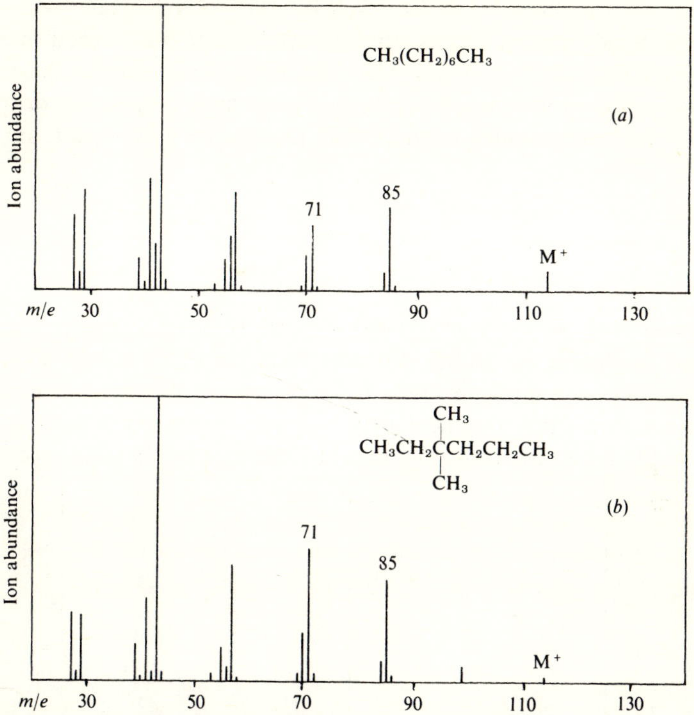

Figure 4.10. Comparison of mass spectra of (*a*) a straight-chain hydrocarbon, n-octane with (*b*) its branched-chain isomer, 3,3-dimethylhexane. Note enhanced abundance of ions at *m/e* 71, 85 in (*b*).

Cycloalkanes give spectra very similar to those of linear alkanes except that the molecular ions are more intense. Once a cycloalkane ring has broken, the residual ion behaves like the ions of linear alkanes, again giving bunches of ions 14 mass units apart; for cycloalkanes, principal fragment ions occur at *m/e* $C_nH_{2n+1-2r}$ where *r* is the number of rings.

The introduction of a double bond into an alkane lowers the ionization potential but does not alter the appearance of the mass spectrum greatly apart from an apparent preferred cleavage at allylic positions as shown in equation (4.6). The latter cleavage has been ascribed to

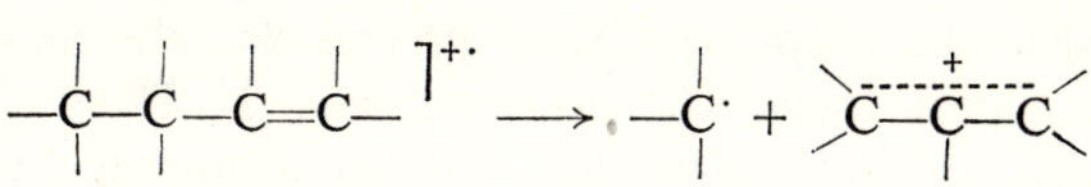

(4.6)

resonance stabilization of the resulting ion. A cycloalkene with the double bond in a six-membered ring exhibits the mass spectrometric 'analogue' of the retro-Diels–Alder reaction. The molecular ion fragments in accord with scheme (4.7) the charge appearing on either fragment. Thermodynamics suggests that the charge will appear on the

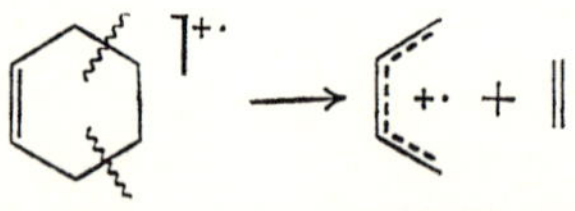

(4.7)

fragment with the lower ionization potential. If the ionization potentials of the fragments are widely different then one of them will take the charge almost exclusively, but if the ionization potentials are close together then either fragment may take the charge and appear as an ion in the mass spectrum. Figure 4.11 illustrates in the mass spectrum

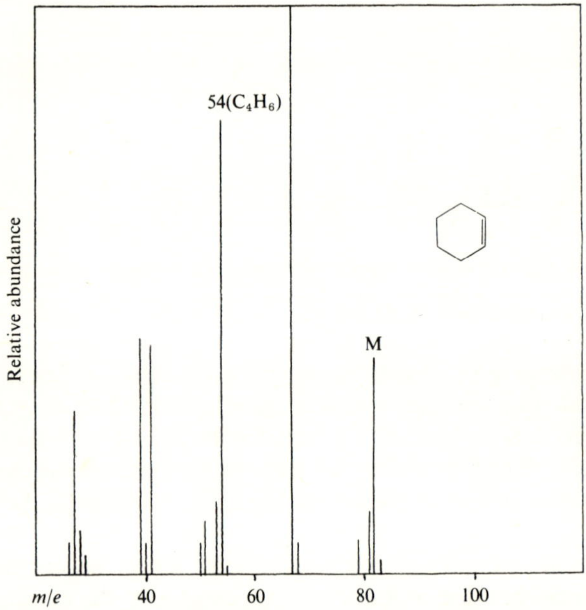

Figure 4.11. Mass spectrum of cyclohexene showing abundant molecular ion peak, and the ion at *m/e* 54 due to a retro-Diels–Alder reaction.

of cyclohexene the retro-Diels–Alder process and the more intense molecular ion peak found with cyclic compounds. There is evidence that, after electron-impact ionization, apparent migration of the double

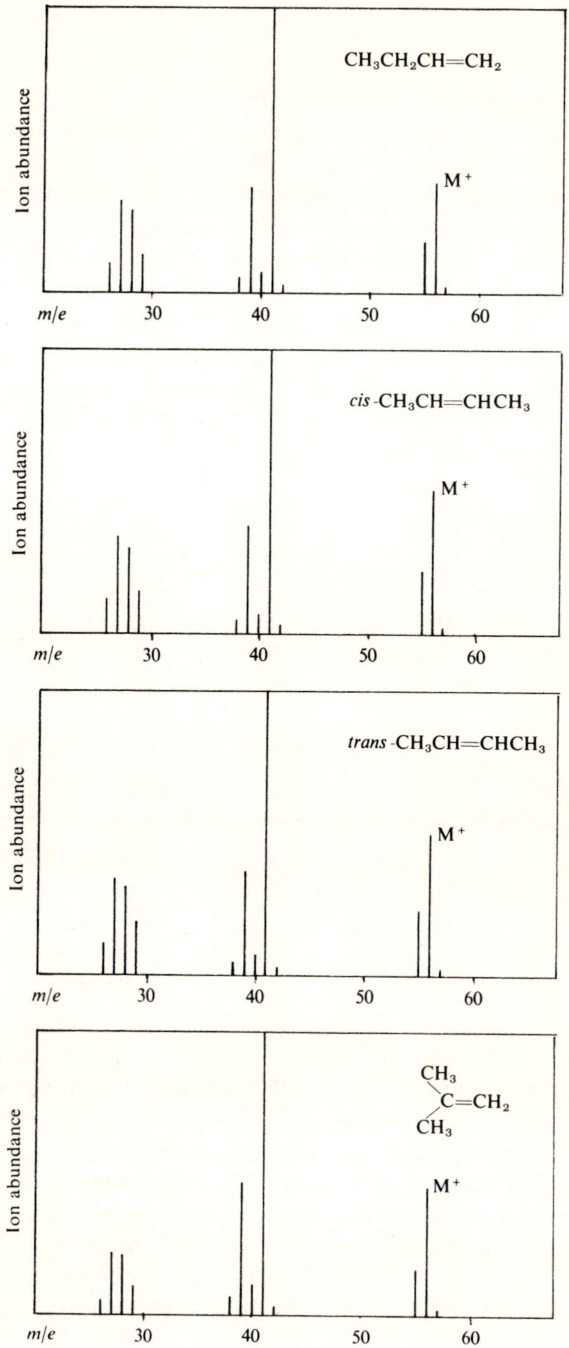

Figure 4.12. Mass spectra of the alkene isomers but-1-ene, *cis*-but-2-ene, *trans*-but-2-ene, and isobutene.

bond occurs in alkenes. The net practical result is that frequently alkene isomers show very similar if not identical mass spectra. Figure 4.12 shows that the isomeric butenes yield very similar mass spectra. Since it can be difficult to reproduce spectra exactly, especially between different laboratories, an identification of any one of these butenes from its mass spectrum alone would be questionable. It is generally not easy to locate a double bond by mass spectrometry on the olefin itself but it may be located by 'fixing' the bond in a derivative. Epoxidation or conversion to an acetonide (4.8) have been recommended as suitable methods for locating double bonds, the heteroatom tending to direct the fragmentation (see following section on heteroatom compounds) as shown. The ions (1) and (2) define the position of the double bond in the original olefin.

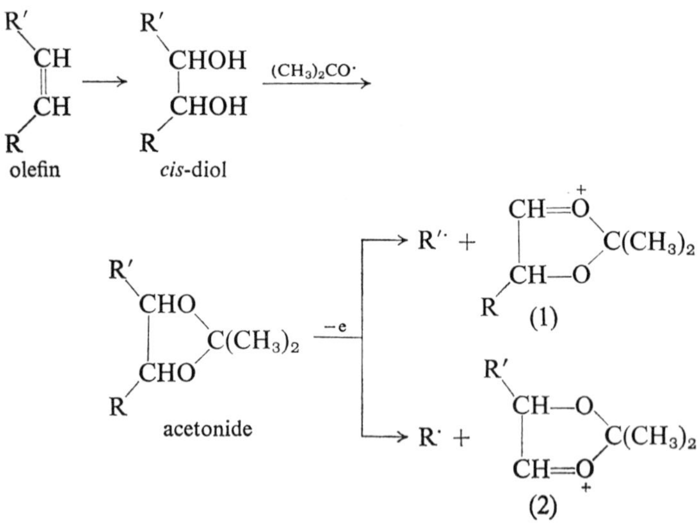

(4.8)

With increasing numbers of double bonds, alkenes give more intense molecular ion peaks and show less of the 'random' bond cleavage characteristic of alkanes and simple alkenes. Terpenoid molecules frequently fragment by loss of C_5-units or multiples of these. The long-chain terpenoid alcohol, solanesol, breaks down in this way in the mass spectrometer, the favoured positions of bond breaking being allylic (4.9). The elimination of 43 mass units (C_3H_7) is commonly observed in terpenoid compounds containing isopropyl, isopropenyl, and iso-

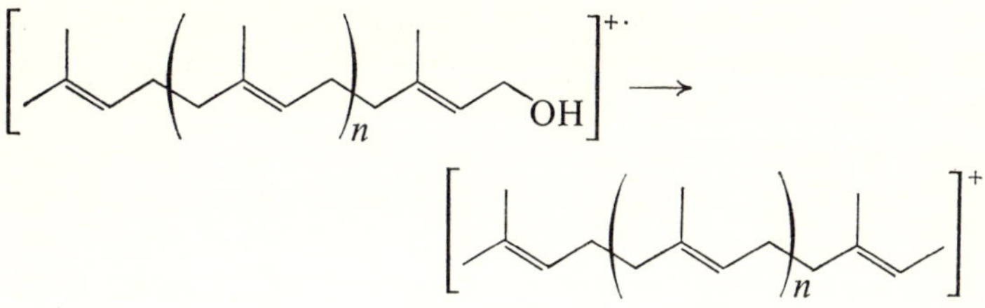

$n = 0–7$

(4.9)

propylidene groups; in the latter two cases, double bond migration must occur.

It was mentioned earlier that the mass spectra of aromatic molecules are characterized by few fragment ions and intense molecular ions. Benzenoid hydrocarbons very often afford an ion at *m*/*e* 91 (C_7H_7) which can be very abundant, particularly where benzylic groups are present (4.10). The $C_7H_7^+$ ion is generally considered to have the

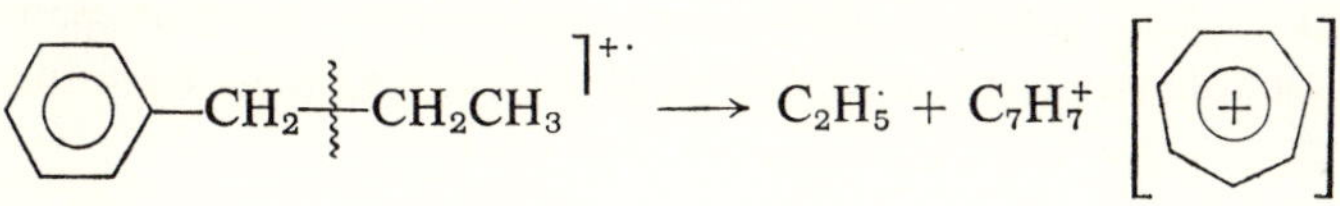

(4.10)

structure of the tropylium ion. It is of value to examine the mass spectrum of a suspected aromatic compound for an ion at *m*/*e* 91 but it should not be assumed that its presence necessarily implies a $C_6H_5CH_2$ unit in the original molecule. The ion at *m*/*e* 91 is frequently observed in the mass spectra of compounds which do not possess a benzyl group; t-butylbenzene, 1,4-diphenylbutadiene, and camphene (4.11) each give an intense peak at *m*/*e* 91 which must be due to rearrangement

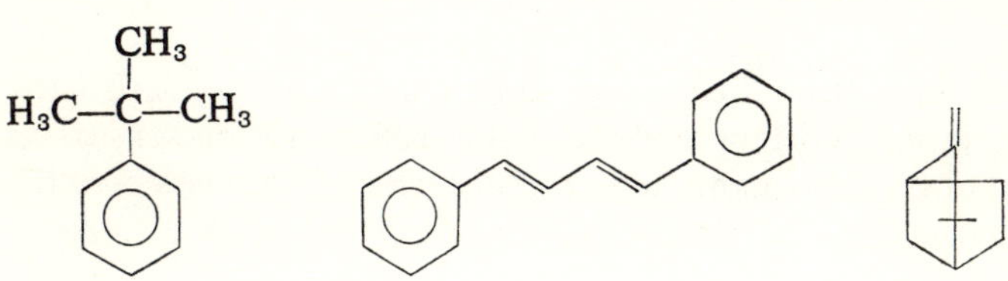

(4.11)

processes. However, an ion at *m*/*e* 91 of composition C_7H_7 is usually a good guide to the aromatic or benzenoid nature of a molecule. Rearrangement processes in the mass spectra of aromatic compounds are often more obvious and widespread than with aliphatic compounds. The presence of heteroatoms frequently makes these reactions more apparent than in hydrocarbons. The consecutive losses of two CO entities from anthraquinone is an example of one such rearrangement (see §4.9 for other examples).

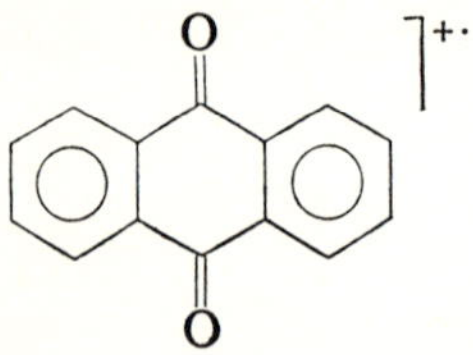

$$\xrightarrow{-CO} C_{13}H_8O^{+\cdot} \xrightarrow{-CO} C_{12}H_8^{+\cdot}$$

4.6. Primary fragmentations of aliphatic heteroatom compounds. The effects of heteroatoms on the mass spectra of organic molecules may be loosely grouped by either regarding the heteroatom as a simple 'saturated' substituent in an alkane or as an 'unsaturated' substituent. Saturated substituents include the halogens, hydroxyl, amino, and sulphydryl and are characterized by having lone-pair electrons. Unsaturated substituents include carbonyl and cyano and have π- and lone-pair electrons. Double bonds and aromatic rings may also be included in the latter category.

TABLE 4.3 *Common 'saturated' substituents* (X) *in* R—X *compounds* (R = alkyl)

F	SH
Cl	SR
Br	NH_2
I	NHR
OH	NR_2
OR	MR_n (M = metal atom)

Saturated substituent heteroatom compounds may be designated R—X, the group X being any one of the commoner ones listed in table 4.3. Some of the more important features of the mass spectrometric fragmentation of these compounds can be grouped under a few headings:

(i) Simple bond cleavage adjacent to the heteroatom is observed with the charge remaining on the hydrocarbon fragment:

$$R \wr X]^{+\cdot} \longrightarrow R^+ + X^\cdot$$

The mass spectra of n-propyl chloride and bromide (figure 4.13) show prominent losses of halogen. Note the change in the isotope pattern between the molecular ion and the first fragment ion containing no halogen. The loss of halogen is so obvious from the isotope patterns and the number of mass units lost (35, 37 for Cl; 79, 81 for Br) that accurate mass measurement is unnecessary.

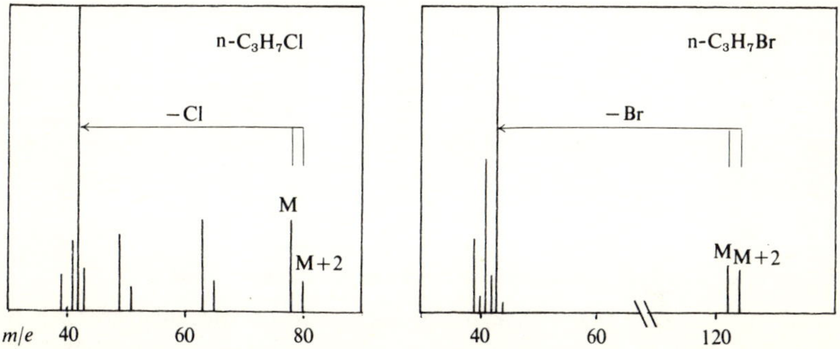

Figure 4.13. Mass spectra of n-propyl chloride and n-propyl bromide. In each spectrum, elimination of the halogen changes the isotope distribution considerably. Peaks above *m*/*e* 45 in the spectrum of n-propyl chloride have been scaled up five times.

The carbon–halogen bond strengths in the molecules, CH_3—X (X = halogen; table 4.4) decrease from the very strong C—F bond to the relatively weak C—I bond. However, in the ions, $CH_3—X^{+\cdot}$, the bond strengths are greatly reduced for X = F, Cl, Br but for the iodo-compound, the bond strength in the ion is greater than in the molecule. The C—Cl, C—Br bonds in the ions are relatively weak and loss of Cl, Br from the molecular ions of organic chlorides and bromides is generally observed. Although the C—F bond strength is similar to that of C—Cl, C—Br in the ion, simple loss of fluorine from organic fluorides is not commonly observed because of the ease of competing reactions (see p. 79 and table 4.5 later).

Table 4.4 gives the bond strengths for several CH_3—X compounds. The order of bond strengths in the molecule are F > OH > Cl > NH_2 > SMe > SH, Br > I but in the molecular ion this order changes to SMe > NH_2 > SH > OMe > OH > I > Br > Cl > F (see also Reed, 1962). Note particularly how the C—I, C—S, and C—N bond strengths increase in going from the molecule to the molecular ion, due possibly to the fact that iodine, sulphur, and nitrogen can stabilize a positive charge much better than fluorine, chlorine, bromine, and

TABLE 4.4 *Relative bond strengths in some* CH_3X *molecules and molecular ions for the processes,* $CH_3X \rightarrow CH_3^{\cdot} + X^{\cdot}$ *and* $CH_3X^{+\cdot} \rightarrow CH_3^+ + X^{\cdot}$; *Bond strengths in kcal/mole and, in parentheses, kJ/mole*

X	D_{CH_3-X}	$D_{CH_3-X^{+\cdot}}$
F	118 (494)	48 (201)
Cl	81 (339)	49 (205)
Br	67 (280)	51 (213)
I	53 (222)	61 (255)
OH	90 (377)	67 (280)
OCH_3	77 (322)	74 (310)
SH	67 (280)	77 (322)
SCH_3	75 (314)	102 (427)
NH_2	80 (335)	91 (381)

oxygen. Unlike the other halogens, iodo-compounds frequently fragment with the iodine taking the charge to give an ion at *m/e* 127.

Elimination of OH, SH, NH_2, from alcohols, thiols, and amines is observed to a slight extent but, particularly with amines, other processes are more important. Cleavage of metal–carbon bonds in simple organometallic compounds is frequently so easy as to lead to low abundances of the molecular ions. One of the most abundant ions in the mass spectrum of tetraethylgermanium is produced by elimination of an ethyl radical from the molecular ion (figure 4.14),

$$[(C_2H_5)_4Ge]^{+\cdot} \xrightarrow{-C_2H_5^{\cdot}} [(C_2H_5)_3Ge]^+ \xrightarrow{-C_2H_4} [(C_2H_5)_2GeH]^+$$
$$m/e\ 190 \qquad\qquad m/e\ 161 \qquad\qquad m/e\ 133$$

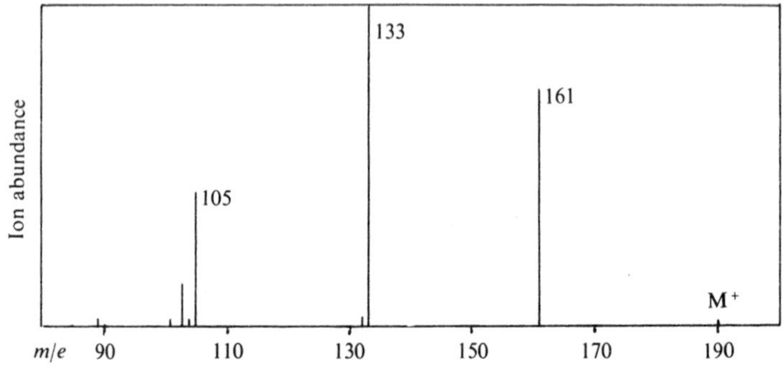

Figure 4.14. Partial mass spectrum of tetraethylgermanium showing large abundance of [M – C_2H_5] ion at *m/e* 161.

(ii) Simple fission adjacent to the heteroatom with simultaneous hydrogen transfer is commonly observed for R—X compounds. Simple cleavage of a molecular ion gives an even-electron species, but cleavage with hydrogen transfer involves more than one bond breaking and gives odd-electron ions. Whereas loss of F from fluoro-compounds is not generally observed, the loss of HF is more common perhaps

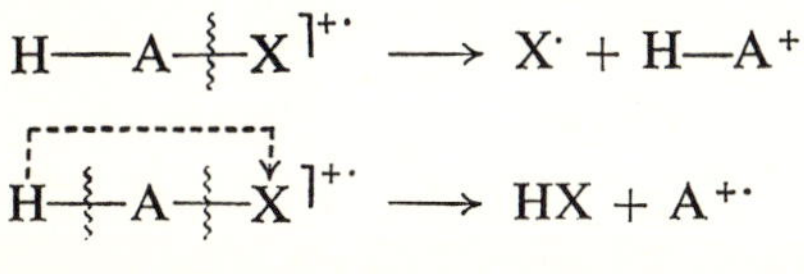

(Even-electron species)

(Odd-electron species)

because of its particularly favourable negative heat of formation. The The elimination of HCl from chloro-compounds is more favoured than the elimination of HBr from bromo-compounds.

Although the ejection of OH from alcohols can be observed in their mass spectra, the elimination of H_2O is usually more common, especially when a suitable hydrogen donor site is available. Deuterium labelling has shown that loss of H_2O from alkyl alcohols predominantly involves a hydrogen on a γ- or δ-carbon atom. A similar process

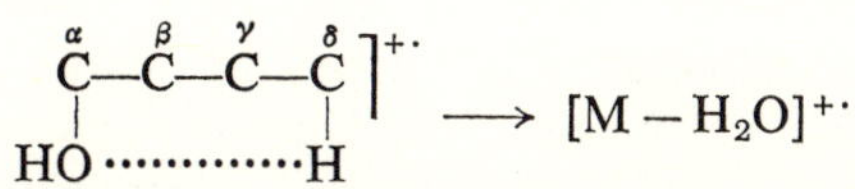

is observed in thiols which eject H_2S, but amines do not readily eliminate NH_3 possibly due to the stability of the ammonium ion making an alternative fragmentation route more favourable (see below, p. 77).

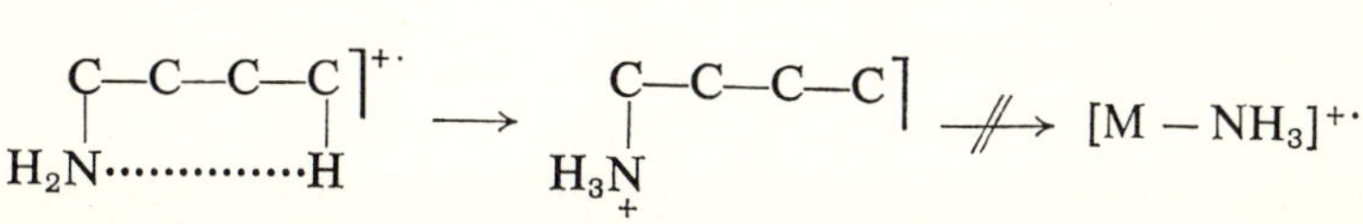

Because of the extreme ease with which H_2O can be ejected from alcohols after ionization, their molecular ions are usually of low abundance. In such cases, field-ionization and chemical-ionization methods are extremely advantageous since these techniques normally yield very good molecular ions. Thus, by comparing electron-impact spectra with those obtained by either field- or chemical-ionization, details of the molecular ion region become easier to interpret. The transfer of hydrogen to hydroxyl during the elimination of H_2O from alcohols seems to involve mainly a six-centre transition state. Deuterium labelling of halogen compounds suggests that a five-membered transition state is most likely during the elimination of hydrohalide. The elimination of ROH from ethers, i.e. cleavage with hydrogen transfer, is not abundant perhaps because of the greater ease of alternative reactions as indicated below. The corresponding reaction in organometallic compounds

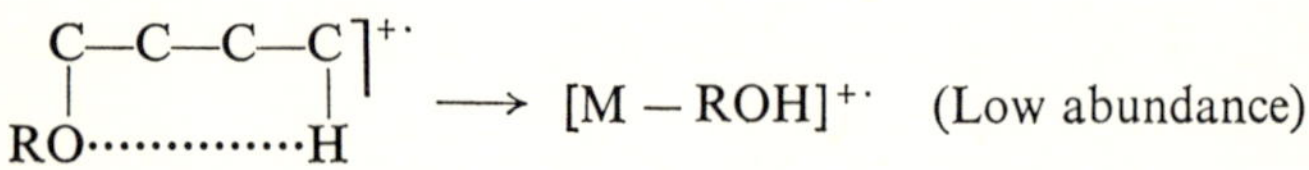

affords (metal + H) ions and these are generally found in their mass spectra.

(iii) Simple cleavage of groups, other than X, from the α-position (α-cleavage) is a common reaction of saturated substituent compounds, R—X. The ejected radical species may be hydrogen or an alkyl radical.

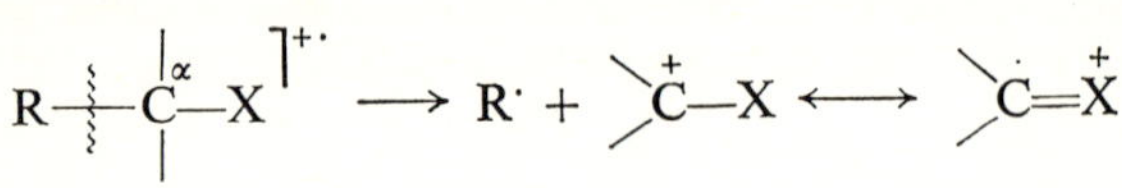

It is thought that stabilization of the positive charge by the group X makes the process favourable. This type of fragmentation is more prominent in those compounds having a substituent (X) which can more readily stabilize the charge, as with the NH_2 group. It was shown by deuterium labelling that elimination of hydrogen from methanol takes place from the carbon atom and not oxygen. The same process

$$\mathrm{H}\!\wr\!\mathrm{CH_2{-}OH}]^{+\cdot} \longrightarrow \mathrm{H^{\cdot}} + \overset{+}{\mathrm{C}}\mathrm{H_2{-}OH} \longleftrightarrow \mathrm{CH_2{=}\overset{+}{O}H}$$

may be observed in ethers, although elimination of an alkyl radical is normally more prominent. The radical may be ejected from either

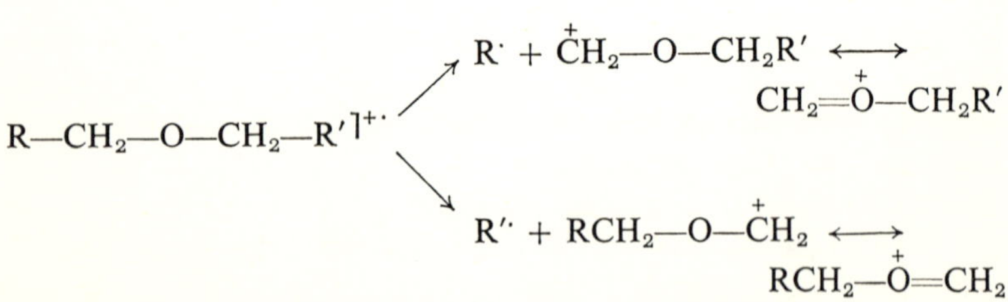

side of the oxygen atom in ethers; the largest group (R, R′) appears to be ejected most favourably. For example, the loss of a propyl radical from ethyl n-butyl ether is more in evidence than the corresponding loss of a methyl radical (figure 4.15). However, examination of the mass

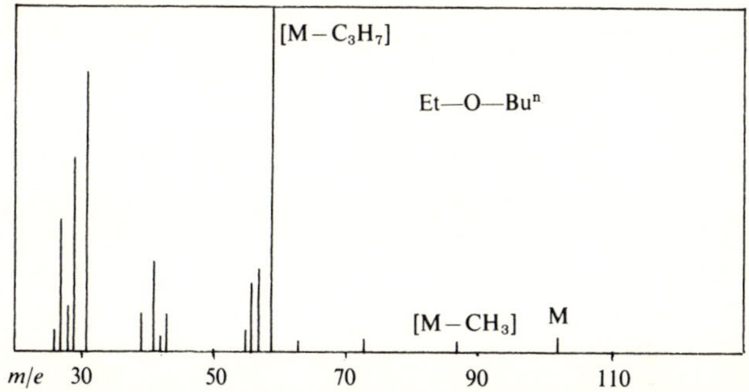

Figure 4.15. Mass spectrum of ethyl n-butyl ether obtained at 70 eV. The peak at *m/e* 59 corresponding to loss of propyl from the molecular ion is much greater than that at *m/e* 87 corresponding to loss of methyl.

spectra of ethers at reduced electron-impact energies suggests that the favourable loss of the larger alkyl fragment may be more apparent than real. At low electron-impact energies, subsequent decomposition of the initially produced fragment ions is reduced and the results then suggest that the smaller alkyl fragment is lost more readily. It is a timely reminder of the importance of the initial internal energy in the molecular ion in determining subsequent fragmentation rates (§§ 1.1 and 3.1) and so determining the abundances of fragment ions.

Similar α-cleavage reactions are found in thioethers and are most marked in amines in which the nitrogen atom can stabilize the positive charge as an immonium species.

$$[CH_3CH_2N(CH_3)_2]^{+\cdot} \longrightarrow$$

$$CH_3^{\cdot} + \overset{+}{C}H_2\text{—}N(CH_3)_2 \longleftrightarrow CH_2\text{=}\overset{+}{N}(CH_3)_2$$

The two types of simple bond cleavage discussed in (i) and (iii) both involve fission adjacent to the α-carbon atom and are competitive processes. In table 4.5 are listed the dissociation energies for α-cleavage of C—C, C—H, and C—X bonds in ions of the type $C_2H_5X^{+\cdot}$ (X = H, F, Br, OH, NH_2, SH, OCH_3, SCH_3, CN). For two competing simple bond fission reactions in the ground state ion, the quasi-equilibrium

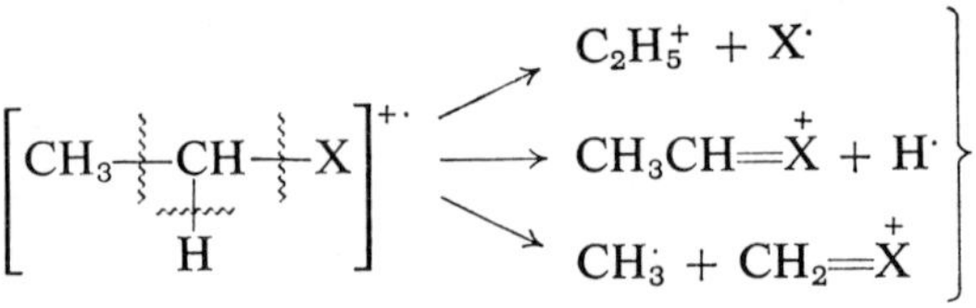

TABLE 4.5 *Bond dissociation energies in ions* $C_2H_5X^{+\cdot}$ *(kcal/mole)*

Ion from	C—C	C—X	C—H
C_2H_6	44	26	26
C_2H_5F	17	29	11
C_2H_5Br	39	30	45
C_2H_5OH	24	49	24
C_2H_5SH	52	56	60
$C_2H_5NH_2$	31	66	29
C_2H_5CN	48	36	56
$C_2H_5OCH_3$	30	53	35
$C_2H_5SCH_3$	59	74	68

theory suggests that their relative rates of fission (k_1, k_2) for any internal energy (E) are given by

$$k_1/k_2 = (\nu_1/\nu_2)\left[\left(1 - \frac{E_1}{E}\right)\Big/\left(1 - \frac{E_2}{E}\right)\right]^{N-1}$$

where ν_1, ν_2 are the frequency factors and E_1, E_2 the energies of activation (§3.3). Approximating this expression through the binomial expansion and putting $\nu_1 \approx \nu_2$ (for simple fission processes, the frequency factors may be equated with the bond vibration frequencies which, for the reactions considered here, are not likely to be sufficiently different to upset the main argument developed by assuming their equality) gives,

$$\ln\,(k_1/k_2) \approx \left(\frac{N-1}{E}\right)(E_2 - E_1)$$

i.e. for two simple cleavage processes from a molecular ion, the relative reaction rates leading to initial fragment ions are proportional to the difference between the energies of activation for the two processes. In the absence of significant activation energies for the reverse reactions, the energies E_1, E_2 may be equated (§3.4) to the ion bond dissociation energies as listed in table 4.5. Hence, inspection of this table should yield at least a qualitative guide to the expected mass spectrometric fragmentations of compounds C_2H_5X i.e. the bond strength differences when equated with $E_2 - E_1$ should give a semi-quantitative measure of ion abundances in their mass spectra. For comparison, the mass spectra of these compounds are recorded in figure 4.16. Striking relationships may be observed immediately and some of these are annotated here:

(*a*) From table 4.5 it can be seen that all the bond dissociation energies are low for $C_2H_5F^+$ and that C—H bond cleavage is particularly easy. It might be expected therefore that, with all bond dissociation energies being low, fluoroethane would not show an abundant molecular ion peak and that the [M – 1] peak would be most prominent. The mass spectrum shows this is the case and confirms that loss of F is a weak process in keeping with the C—F bond being the strongest of a set of weak ones.

(b) By way of contrast to fluoroethane, in bromoethane all the bonds are moderately strong with the C—X (X = Br) the weakest, the C—C bond stronger, and the C—H bond the strongest. It might be expected therefore that a good molecular ion would be observed and the most prominent process the loss of Br. The mass spectrum is in keeping with this deduction.

(*c*) The mass spectrum of ethane thiol, C_2H_5SH, shows a large molecular ion peak with C—C and C—SH bond cleavages of similar importance. Table 4.5 shows that the bonds in the $C_2H_5SH^+$ ion are all rather strong and that C—C, C—SH α-cleavage reactions are almost equally easy.

(*d*) In ethanol and ethylamine, the bond dissociation energies for C—C and C—H fissions are very much less than for C—X fission. These substances should give, therefore, moderately abundant molecular ions with marked losses of hydrogen and methyl radicals. Examination of their mass spectra bears out this argument.

Table 4.5 may be used to 'predict' the mass spectra of the simple compounds C_2H_5X. Because the compounds are so simple further decomposition of the initial fragment ions is greatly reduced. The abundance of an ion in a mass spectrum depends upon its rate of formation and upon its rate of decomposition (§1.1) but in the case of C_2H_5X compounds further decomposition of the initial fragment ions is negligible and their rates of formation determine their abundances. The

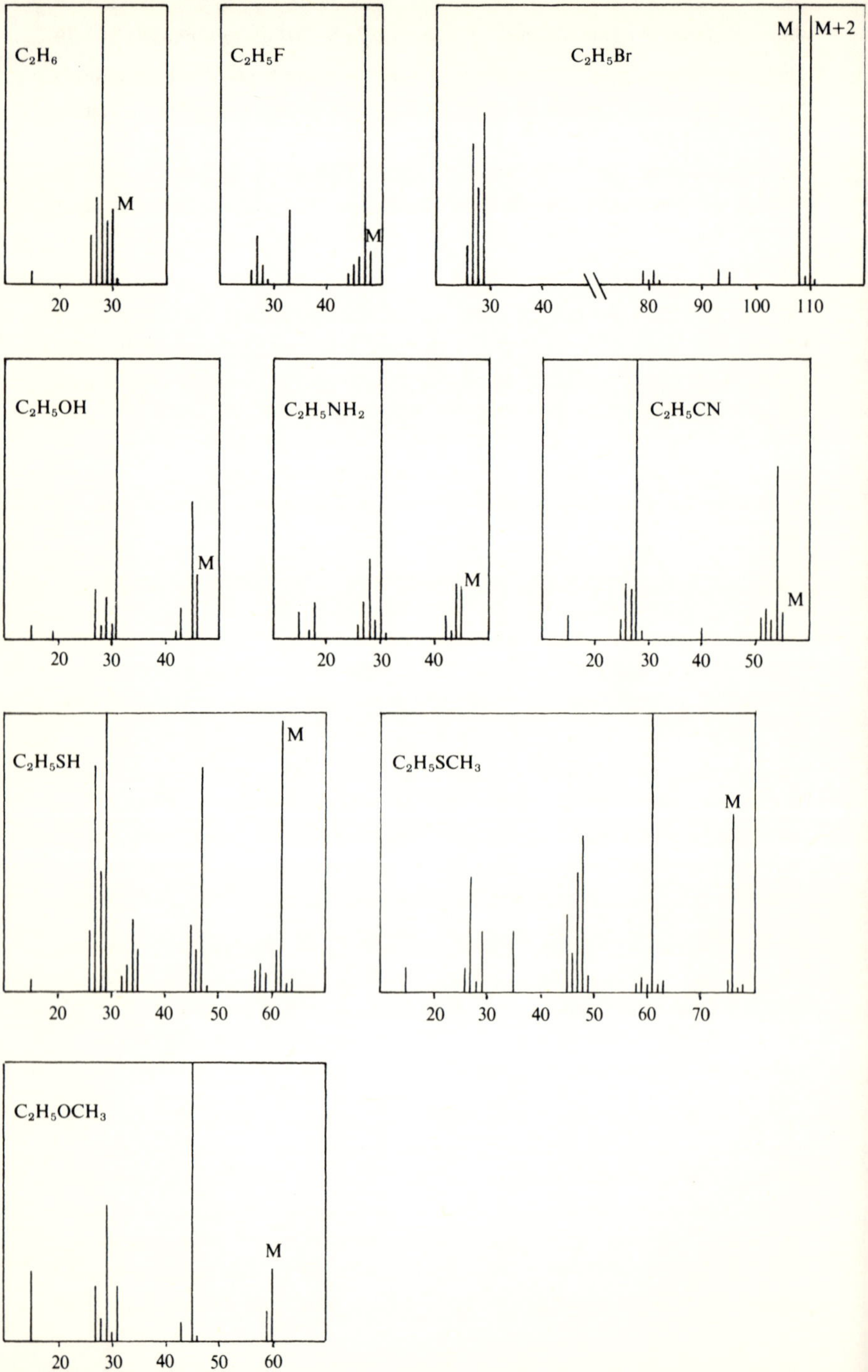

Figure 4.16. Mass spectra of compounds C_2H_5X (X = H, F, Br, OH, NH_2, CN, SH, SCH_3, OCH_3).

general principles apparent from an examination of table 4.5 carry through to a large extent in more complex molecules and serve to provide a thermochemical understanding of many fragmentation processes. In the remainder of this section on fragmentation reactions, reference to the bond dissociation energies given in table 4.5 will prove of value.

It is interesting to note that the C—C bond dissociation energy in ethane is comparatively large and greater than for C—H bond fission; these relative strengths are reflected in the appearance of the mass spectrum which exhibits a strong loss of hydrogen from the molecular ion. However, as the alkyl chain length increases in alkanes, C—C bond dissociation energy decreases more rapidly than C—H bond dissociation energy. Table 4.6 illustrates this effect for loss of a methyl

TABLE 4.6 C—C *bond dissociation energies in* n–*alkane ions*†

Ion from	D_{C-C}
C_2H_6	44 (184)
C_3H_8	30 (126)
C_4H_{10}	21 (88)
C_5H_{12}	15 (63)

† Calculated for reaction, $C_nH_{2n+1}\cdot CH_3^{+\cdot} \rightarrow C_nH_{2n+1}^{+} + CH_3^{\cdot}$, in kcal/mole and, in parentheses, kJ/mole.

radical from n-alkane molecular ions; for n-butane the relative strengths of C—H and C—C bonds are very similar and much less than the values

$$[C_nH_{2n+1}CH_3]^{+\cdot} \longrightarrow [C_nH_{2n+1}]^{+} + CH_3^{\cdot}$$

in ethane and most of the other compounds in table 4.5. In long-chain alkyl compounds, $C_nH_{2n+1}X$, this reduction of the C—C bond strength, coupled with the greater statistical probability of the excess of energy in the ion appearing in a C—C bond, may be expected to lead to an increasing importance of general fragmentation of the alkyl chain; such an effect is observed in the mass spectra of these compounds. As the chain length increases, the specific fragmentations associated with the heteroatom, X, decrease in importance and extensive fission of the alkyl chain is observed.

(iv) Bond breaking at positions other than adjacent to the α-carbon atom is observed in the mass spectra of aliphatic heteroatom compounds. As the proportion of heteroatom in a compound decreases, these alternative bond breaking processes become more prominent, and α-cleavage of less importance. Thus, in the long-chain compound shown in scheme (4.12) cleavages such as these at the positions γ, δ, ... increase in importance as R increases in length.

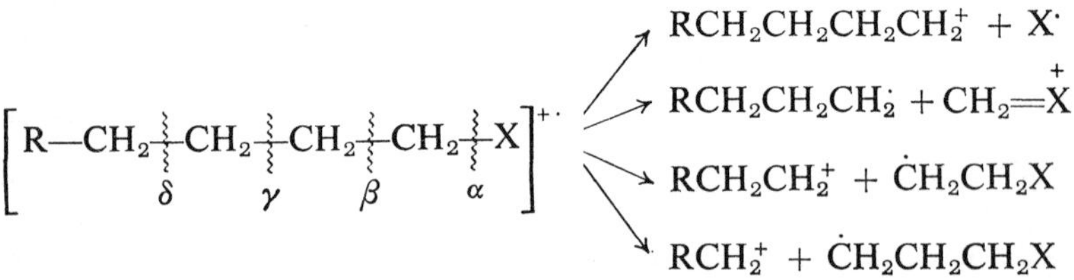

(4.12)

In alkyl chlorides and bromides, $C_nH_{2n+1}X$, this simple bond breaking reaction may also lead to abundant ions, $C_nH_{2n}X$, particularly for $n = 3, 4, 5$. The propensity for this reaction shown by these compounds has been attributed to the formation of cyclic structures similar to those found in solution chemistry. The abundance of ions formed by this process decreases with chain branching in the alkyl moiety and the reaction is not prominent in fluorine or iodine compounds.

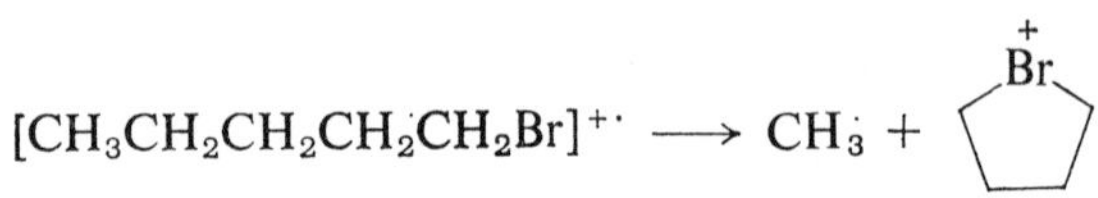

A similar increased abundance of $C_nH_{2n}X$ ions at $n = 3, 4, 5$ is found in amines but not alcohols or thiols. Unlike the halogen compounds, amines give more abundant ions at $n = 5$, which has been ascribed to a preference for six-membered ring formation (compare losses of HX from R—X compounds where five- and six-membered transition states appear to be involved).

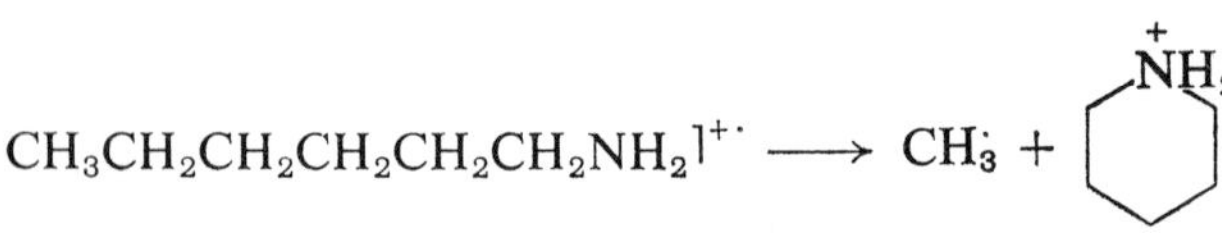

The commonest 'unsaturated' substituents of alkanes are the carbonyl functional groups >CO, —CHO, $—CO_2H$, $—CO_2R$, $—CONH_2$, and the sulphoxide, sulphone, phosphate, and cyano-groupings.

The mass spectra of ketones are dominated by bond breaking adjacent to the α-carbon atom (α-cleavage) similar to reaction scheme (iii) shown above for saturated substituents. The comparable breaking of the carbon to heteroatom bond, as in (i) above, is not observed presumably because of the increased strength of C═O bonds as against C—OH.

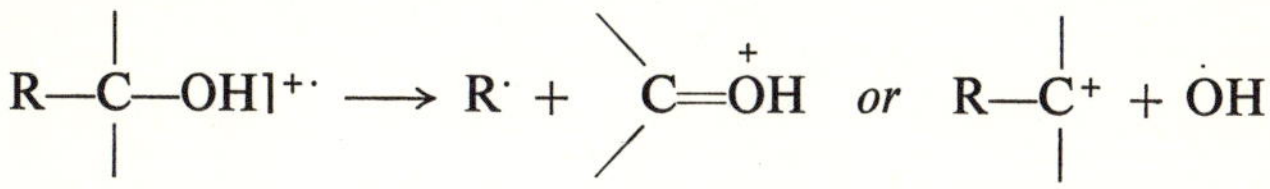

but

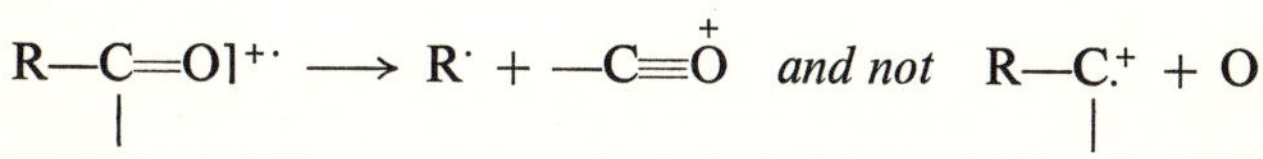

The cleavage adjacent to carbonyl is reminiscent of the photolytic decomposition of ketones and aldehydes. The large degree of resonance stabilization of the positive charge in the fragment ion is said to be responsible for the ease of fission at the α-position. As with α-cleavage of saturated substituent groups, cleavage adjacent to the carbonyl group appears to lead preferentially to the ejection of the largest radical (R, R′) from the molecular ion in 70-eV spectra. The α-cleavage process is frequently followed by elimination of CO from the fragment ion. The mass spectrum of ethyl n-propyl ketone (figure 4.17) shows the elimination of both ethyl and n-propyl groups followed by ejection of CO.

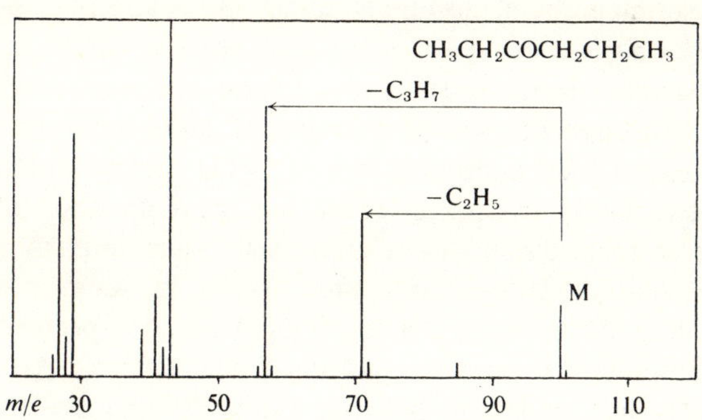

Figure 4.17. Mass spectrum of ethyl n-propyl ketone showing that ion abundance at *m/e* 57 corresponding to loss of C_3H_7 is greater than that at *m/e* 71 corresponding to loss of C_2H_5 at 70 eV.

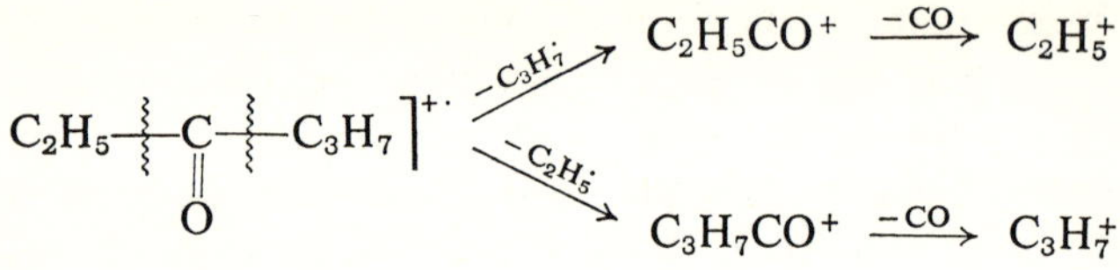

or, in charge localization terms,

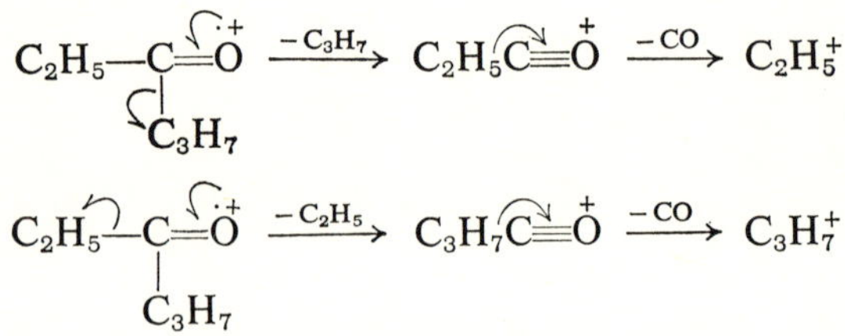

In aldehydes, one of the groups adjacent to the carbonyl is simply hydrogen, and elimination of atomic hydrogen from these compounds is observed.

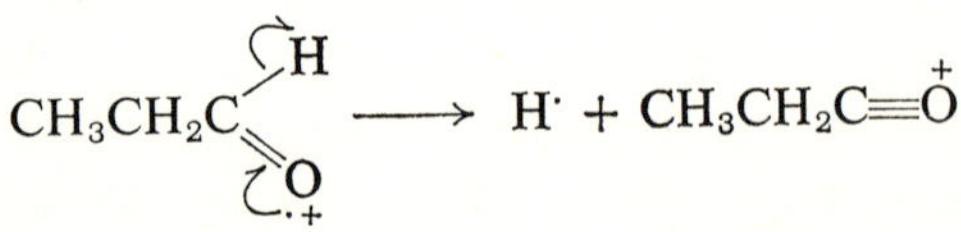

As well as this simple type of bond breaking reaction, ketones and aldehydes exhibit loss of H_2O from their molecular ions. The process has been examined by deuterium labelling and found to be complex.

The α-cleavage adjacent to carbonyl is still important in the mass spectra of acids, esters, and amides, and these functional groups may be recognized through the loss of appropriate units from the molecular ions. In each of the examples shown in (4.13), simple fission at either side of the carbonyl group may occur, but the abundances of the fragment ions so formed can vary a great deal. Thus, aliphatic acids and amides generally produce only small peaks in their mass spectra corresponding to the loss of OH and NR'_2 (4.13*a,c*) respectively whereas the ejection of OR′ from esters (4.13*b*) is much more noticeable; aromatic acids and amides do however eject OH and NR'_2 from their respective molecular ions. Usually, cleavage between the alkyl chain and the carbonyl function is more prominent and $[M-CO_2H]$, $[M-CO_2R']$, $[M-CONR'_2]$ ions are observed. Also, the ejected moieties CO_2H, CO_2R', and $CONR'_2$ appear as charged species at the

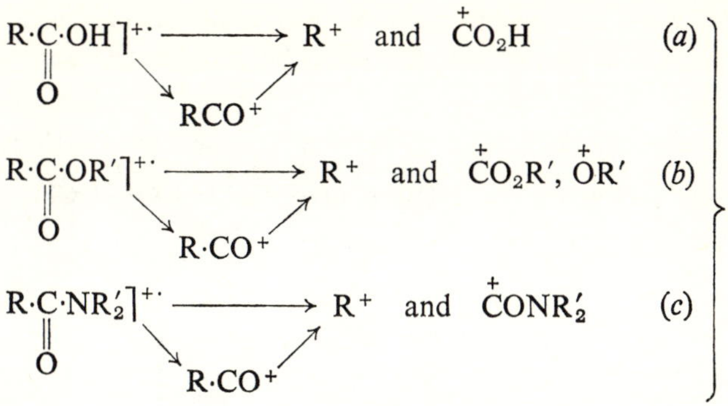

(4.13)

relevant *m/e* values. An aliphatic acid methyl ester, RCO_2CH_3, for example, affords ions at R^+, RCO^+, $CO_2CH_3^+$, and OCH_3^+.

Similar α-cleavages are found in the mass spectra of sulphoxides and sulphones but not in cyanides. Although metal carbonyls, $M(CO)_n$, are not really analogous to organic carbonyl compounds, it is convenient to note here that α-cleavage of the carbon monoxide ligand is extremely easy in these organometallics. Where several carbonyl ligands are present, fragment ions are observed corresponding to successive losses of CO.

$$M(CO)_n^{+\cdot} \longrightarrow M(CO)_{n-1}^{+\cdot} \longrightarrow M(CO)_{n-2}^{+\cdot} \longrightarrow \text{etc.} \longrightarrow M^{+\cdot}$$

The α-cleavage in amides is very important for sequence analysis of amino-acid residues in peptide chains. Bond breaking adjacent to carbonyl yields type A and type B ions as shown in the scheme (4.14). Type A ions may eject CO to give type B ions. A peptide chain normally cleaves at each CO—NH bond to yield type A fragment ions and by

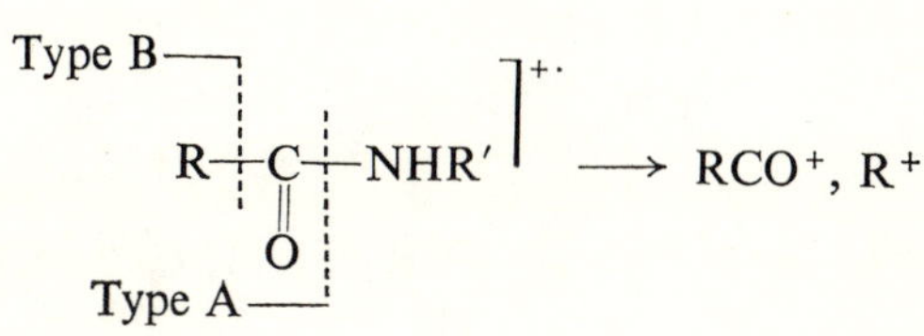

(4.14)

examining these ions, the sequence may be determined from the mass spectrum. It is usual to acetylate the free amino-end of the peptide and to convert the acid terminal to its methyl ester. The tripeptide, Leu·Ala·Gly, methylated and acetylated in this way gives fragment ions at *m/e* 156 ($C_8H_{14}NO_2$), 227 ($C_{11}H_{19}N_2O_3$), 284 ($C_{13}H_{22}N_3O_4$),

315 ($C_{14}H_{25}N_3O_5$; molecular ion) from which the sequence is immediately obtained. For simple peptides it may be unnecessary to determine

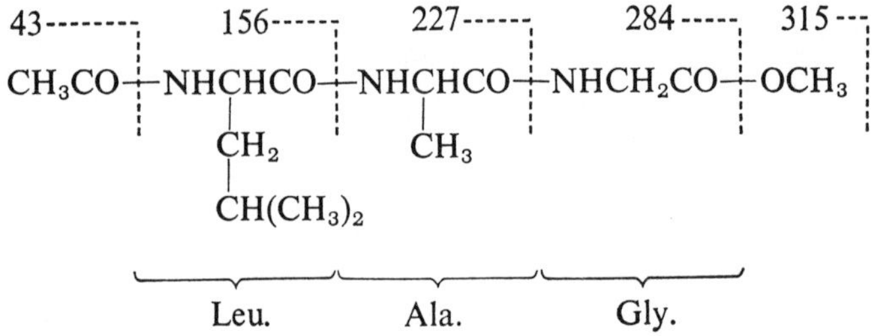

the compositions of the fragment ions except as confirmation of the sequence. Larger peptides often give complex mass spectra because of competing side-chain fragmentation, and it is then necessary to determine the elemental compositions of all ions in the spectrum by accurate mass measurement (§5.2). From these elemental compositions the sequence may be determined as above and computer programs have been developed to deal with the sorting involved.

A further parallel between compounds with saturated and those with unsaturated substituents is found in the six-centre transfer of hydrogen in carbonyl compounds comparable with the six-centre elimination of H_2O from alcohols. The reaction is often called the McLafferty rearrangement and is exemplified in the scheme (4.15) for methyl n-propyl ketone. The elimination of H_2O from n-pentanol is shown for comparison. Again the rearrangement reaction of the odd-electron molecular ion involves more than one bond cleavage and yields an

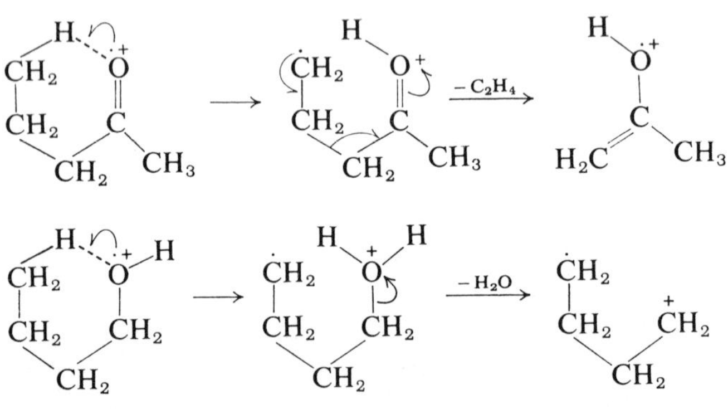

(4.15)

odd-electron fragment ion. In comparing the two reactions (4.15) notice that for the carbonyl compounds, the charge generally remains with the heteroatom fragment and a neutral aliphatic moiety is ejected but for alcohols, the charge remains with the aliphatic part and the heteroatom is ejected as neutral H_2O.

Just as the elimination of H_2O from alcohols is paralleled by the elimination of H_2S from thiols, the six-centre rearrangement of carbonyl compounds may be regarded as a particular instance of the general phenomenon shown schematically in reaction (4.16). The atoms X,Y may be carbon, oxygen, sulphur, phosphorus, or nitrogen in various combinations. Deuterium labelling has demonstrated that

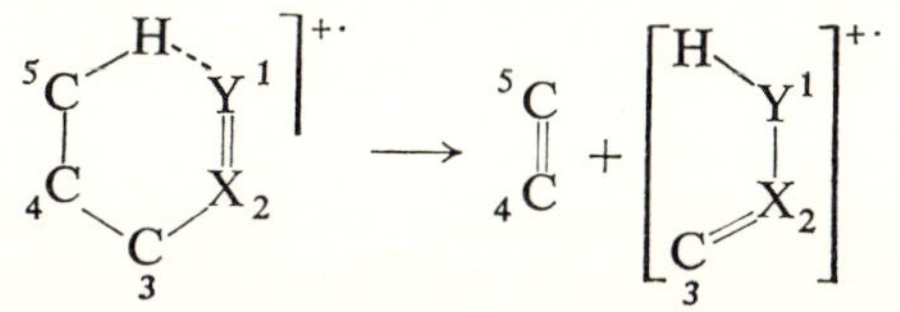

(4.16)

hydrogen transfer takes place mainly from position 5 but transfer also appears to arise in positions 3 and 4. The rearrangement is observed in the mass spectra of amides and sulphoxides, for example as in reaction schemes (4.17).

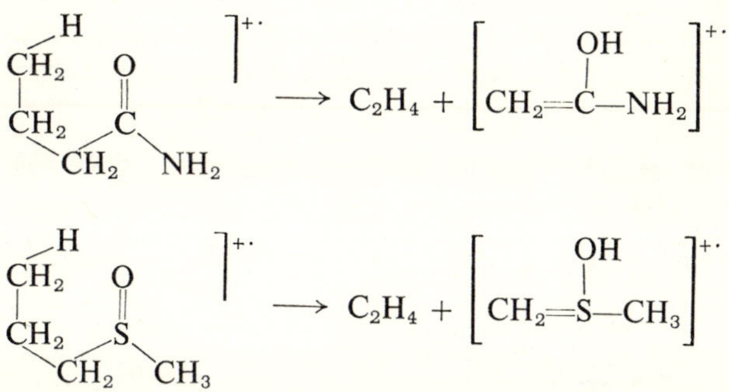

(4.17)

Esters having an alkoxy group longer than methoxyl also show the rearrangement and give a peak corresponding to the ionized free acid.

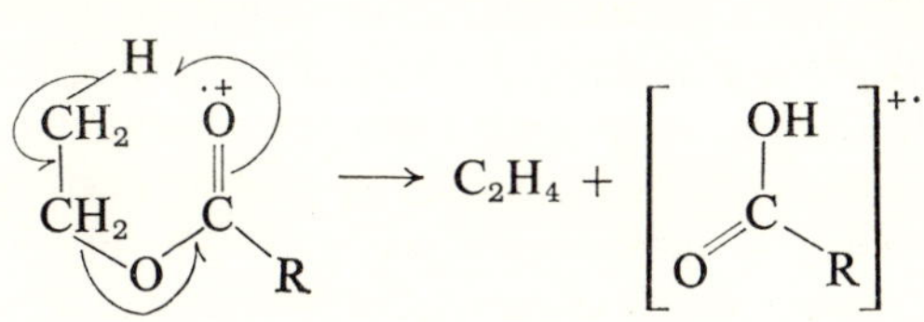

Sometimes this single hydrogen transfer in esters is masked by a double hydrogen transfer to give what is thought to be the protonated acid. The n-propyl ester of acetic acid (figure 4.18) gives a large peak at

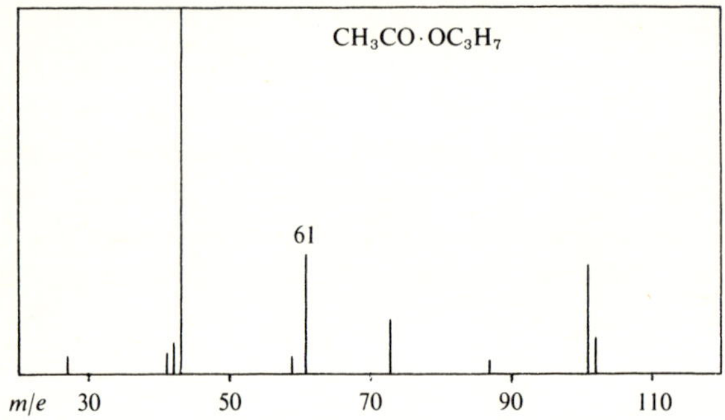

Figure 4.18. Mass spectrum of n-propyl acetate showing peak at *m/e* 61 due to double hydrogen rearrangement. The peaks above *m/e* 100 have been scaled up 100 times.

m/e 61 due to this double rearrangement. Deuterium labelling suggests that the two hydrogen atoms are transferred mainly from the β- and γ-carbon atoms of the alkoxy-group, but the results of the labelling studies leave some ambiguity. A similar double hydrogen rearrangement with formation of a protonated species is found in the mass spectra of many types of esters including carbonates and those of di-acids.

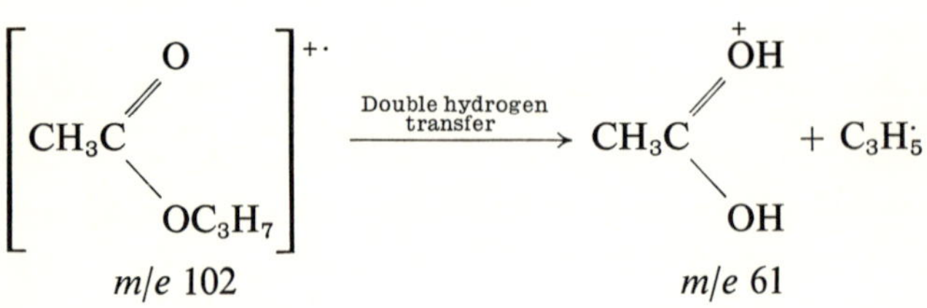

The generalized simple cleavage of aliphatic chains in reaction type (iv) of saturated substituents is found in a modified form with unsaturated substituents in alkanes. Thus, esters of general formula $C_nH_{2n+1}CO_2R$ give rise to fragment ions of composition $C_nH_{2n}CO_2R$. However, unlike halide compounds where the most abundant fragment ions have a composition $C_nH_{2n}X$ with $n = 4$, the ester fragment ions are most abundant at the compositions $n = 2, 6, 10$, etc. The effect has been explained in terms of transfer of hydrogen and fragmentation as shown in the reaction sequence (4.18) but there is little experimental confirmation of this view.

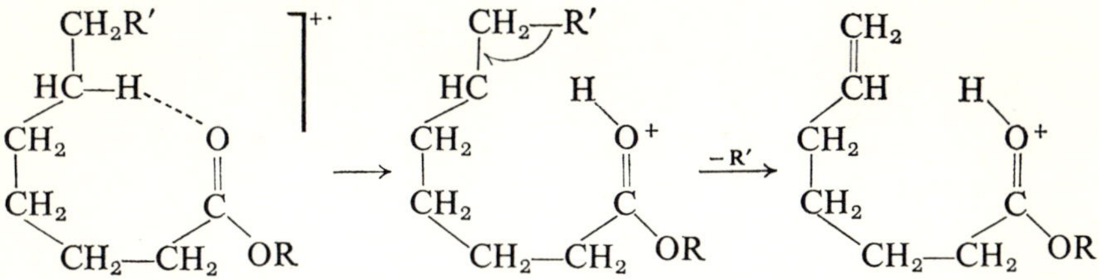

(4.18)

Many simple fragmentation reactions of heteroatom molecular ions may be summarized by reference to scheme (4.19) in which X represents the heteroatom and R^1, R^2, R^3, are substituents; when X is doubly

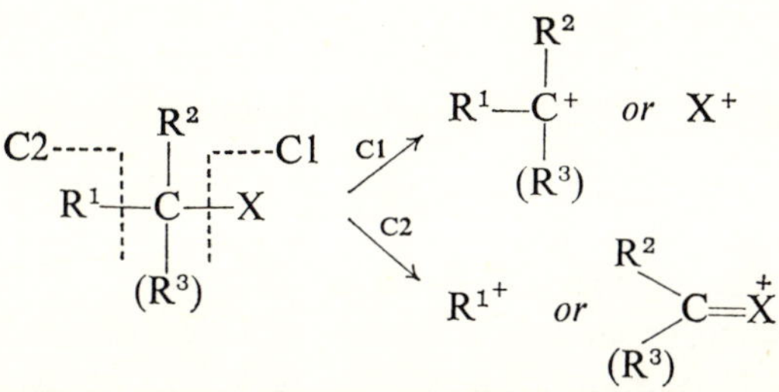

(4.19)

bonded to the adjacent carbon atom then R^3 is not present. Type C1 cleavage is breaking of the C—X bond and is observed with saturated heteroatom substituents (X = Cl, Br, I, OH, SH, NH_2). The C1 cleavage is not observed with unsaturated heteroatom substituents. Type C2 cleavage is commonly observed for both saturated and unsaturated heteroatom substituents of alkanes. The extent to which C1 or C2 bond breaking occurs is very dependent on the nature of the heteroatom and the structure of the molecule. Either or both fragments from C1, C2 reactions may appear as even-electron charged species in the mass spectrum as shown in (4.19).

The other common process observed in aliphatic heteroatom compounds is rearrangement of hydrogen accompanying elimination of a neutral species to give odd-electron fragment ions. Again the process can be represented in a common formalism (4.20) by two reaction types (RE1, 2). With saturated substituents (X), the rearrangement–

elimination reaction (RE1) is general except for amines, but for unsaturated substituents the alternative process (RE2) is general. The processes shown in (4.20) have been drawn with the six-membered

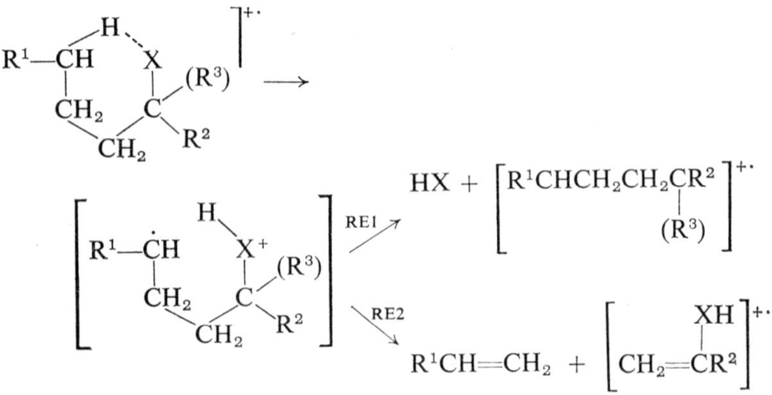

(4.20)

R^3 absent for unsaturated substituent (X).

ring transition state, but it should be recalled that the results of deuterium labelling are not unequivocal. For halogen compounds, a similar five-membered state appears to be more important.

A double bond or an aromatic ring can be considered here as an unsaturated substituent in alkanes. The 'allylic' or 'benzylic' types of cleavage adjacent to these unsaturated centres are comparable with, for example, the α-cleavage (C2) of carbonyl compounds if the unsaturated centres take the place of the heteroatom. The allyl cation is isoelectronic

$$[CH_3 \wr CH{=}O]^{+\cdot} \longrightarrow CH_3^{\cdot} + H\overset{+}{C}{=}O \longleftrightarrow HC{\equiv}\overset{+}{O}$$

$$[CH_3 \wr CH_2{-}CH{=}CH_2]^{+\cdot} \longrightarrow CH_3^{\cdot} + \overset{+}{C}H_2{-}CH{=}CH_2 \longleftrightarrow CH_2{=}CH{-}\overset{+}{C}H_2$$

with the formyl cation in the resonance formalism; in the first case, a positive charge is delocalized over three carbon atoms and in the other, over one carbon and one oxygen atom. The comparison is not a particularly good one however because the olefinic double bond itself may be considered isoelectronic with the chloro-function for example. As with cleavage adjacent to a chlorine atom, cleavage adjacent to the olefin group occurs with hydrogen transfer and is an important process. In many ways, double bonds and aromatic rings have characteristics in common with both saturated and unsaturated substituents.

$$\text{R—Cl}^{+\cdot} \longrightarrow \text{HCl} + [\text{R}-\text{H}]^{+\cdot}$$

$$\text{R—CH}{=}\text{CH}_2^{+\cdot} \longrightarrow \text{H}_2\text{C}{=}\text{CH}_2 + [\text{R}-\text{H}]^{+\cdot}$$

It must be stressed that, so far, only primary fragmentation of the molecular ion has been considered. The effects of subsequent fragmentation may modify the mass spectrum to such an extent that the primary processes appear to have little prominence. For example if the rate of fragmentation of M^+(4.21) to give A^+ is slower than the fragmentation of A^+ to B^+ then the intensity of the peak corresponding to A^+

$$M^+ \longrightarrow A^+ \longrightarrow B^+ \qquad (4.21)$$

in the mass spectrum will be very much reduced (§1.1). Also, subsequent fragmentation is not so easy to classify and will be discussed later. If only for these two reasons there is little to substitute for experience gained in the interpretation of mass spectra. Attempts are being made, with some success, to program computers to elucidate structures given just the mass spectrum. In this respect, the exceptional memory characteristics of a computer are invaluable for storing accumulated knowledge and experience.

4.7. Primary fragmentations of aromatic heteroatom compounds. The mass spectra of aromatic compounds may be subdivided into three classes: those in which the aromatic ring is separated from the heteroatom by an aliphatic part (aralkyl compounds), those in which the aromatic ring is attached directly to the heteroatom, and those in which the heteroatom forms part of the aromatic ring. Since the presence of the aromatic ring may modify the mass spectrum considerably compared with what is observed with aliphatic compounds these three classes are considered separately.

(i) Aralkyl compounds have mass spectra with characteristics common to aliphatic and aromatic compounds separately. The presence of the aromatic ring appears to confer greater stability on the molecular ions which are much more prominent than for simple aliphatic materials. The aromatic ring may itself act as the substituent X in formula (4.19) as mentioned above, and give corresponding fragmentations. Thus, cleavage and rearrangement processes C1, C2, RE2 (4.22*a*,*b*,*c*) are observed in n-butylbenzene.

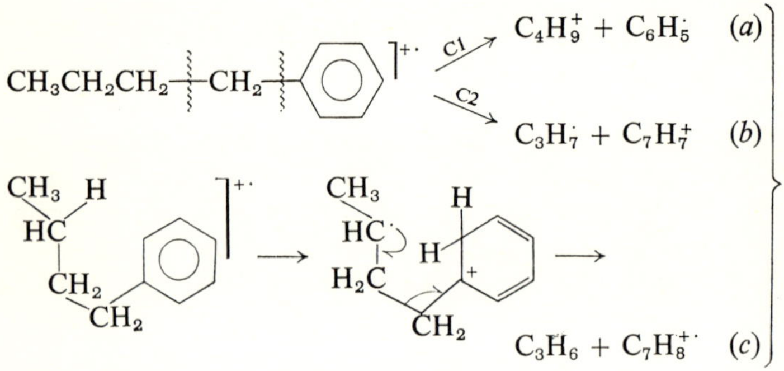

(4.22)

Fragmentation at other positions in the aliphatic chain is also found and for long chains this supersedes the above C1, C2, RE2 reactions in importance. Similar behaviour was discussed earlier for aliphatic heteroatom compounds with increasing chain length.

The presence of the aromatic ring in hetero-compounds in effect introduces a second substituent (X) and, in the absence of interaction between the substituents through space, fragmentation of the molecular ion is influenced by each. Thus, the decomposition of methyl 3-phenylbutyrate (4.23) on electron-impact gives some of its fragment ions as shown. The closer the two substituents to each other, the more they

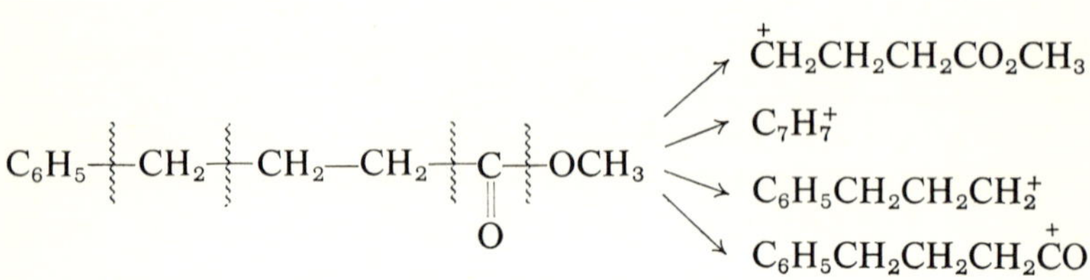

(4.23)

may interact through space or through bonding so as to modify their separate effects.

(ii) When the substituent heteroatom is attached directly to a benzene ring, simple C1 cleavage may still occur as with chlorobenzene and nitrobenzene. It is interesting to note that the C—F bond strength is

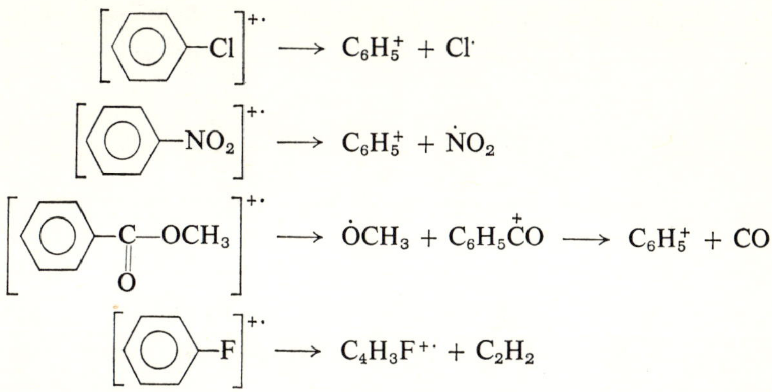

great enough to prevent simple cleavage as also with aliphatic compounds and instead, C_2H_2 is eliminated from the benzene ring. The elimination of halogen from chlorobenzene may be an ion pair process.

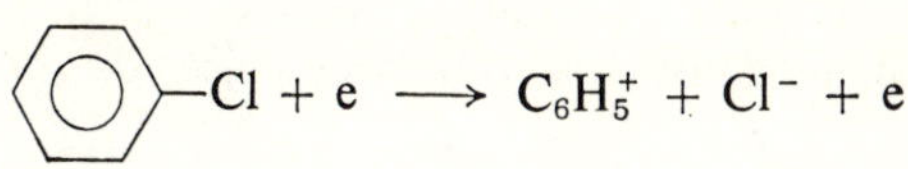

This would be a case of primary cleavage into positive and negative ions on electron-impact and is observed for halogen compounds at low electron-beam energies. Since we are concerned with positive ions and do not 'see' the negative ones in most organic mass spectrometry, the existence of this process does not affect the observed spectrum except in so far as the energy of the process is different from that in which a chlorine atom is ejected (§1.1).

Often, simple Cl cleavage does not occur and rearrangement of the molecular ion takes place before fragmentation. Rearrangement processes are covered in more detail in §4.5 but those germane to this description of the fragmentation of aromatic heteroatom compounds are discussed here.

Phenols, phenyl ethers, and amines commonly occur in aromatic chemistry and exhibit complex fragmentation behaviour. Phenol eliminates not OH, but CO and CHO from the molecular ion. Anisole ejects CH_2O and CHO but not methyl or methoxyl, and aniline eliminates HCN from the molecular ion. These processes are quite general.

In some circumstances, nitro-compounds do not simply eject NO_2 but a complex reaction sequence occurs in which NO and CO are ejected successively. This process is observable to a slight extent in the mass spectrum of nitrobenzene itself but it can increase in importance

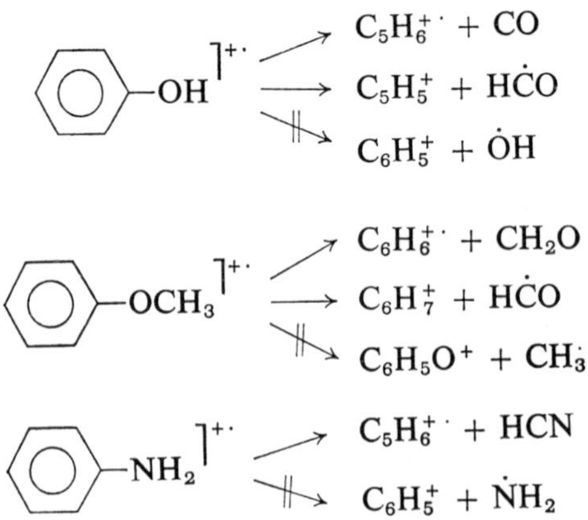

in other compounds so as to make the simple elimination of NO_2 inconsequential. Table 4.7 lists the commoner substituents exhibiting such rearrangements and the neutral species ejected.

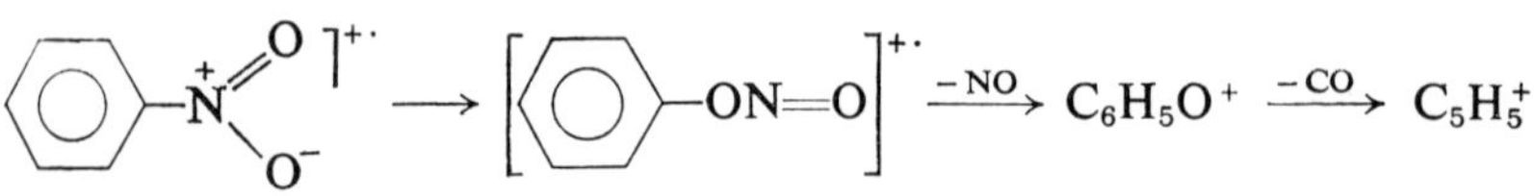

TABLE 4.7 *More commonly occurring substituents of aromatic rings which eject neutral particles resulting from rearrangement*

Substituent	Neutral particle(s) ejected after rearrangement†
NO_2	NO, CO (NO_2)
NH_2	HCN
$NHCOCH_3$	C_2H_2O, HCN
CN	HCN
F	C_2H_2
OCH_3	CH_2O, CHO
OH	CO, CHO
SO_2NH_2	SO_2, HCN
SH	CS, CHS (SH)
SCH_3	CS, CH_2S, SH (CH_3)

†Particles shown in parentheses are ejected without rearrangement. Where these are given, it indicates that these particle losses are usually observed along with those resulting from rearrangement; their relative proportions are very variable.

Often there is more than one heteroatom substituent in an aromatic compound and they may interact to modify the above fragmentation processes considerably. The relative positions of the substituents around the aromatic ring are also important. Whereas the elimination of methyl from the methoxyl group in anisole is a scarcely detectable process, in *ortho*- and *para*-anisidine the process is more important than the usual behaviour of the aromatic methoxyl group (loss of CH_2O, CHO). The ease of elimination of methyl from *o*-, *p*-anisidines has been ascribed to a resonance stabilization of the fragment ion. The behaviour of *m*-anisidine is more 'normal' in that CH_2O and CHO are eliminated from the molecular ion.

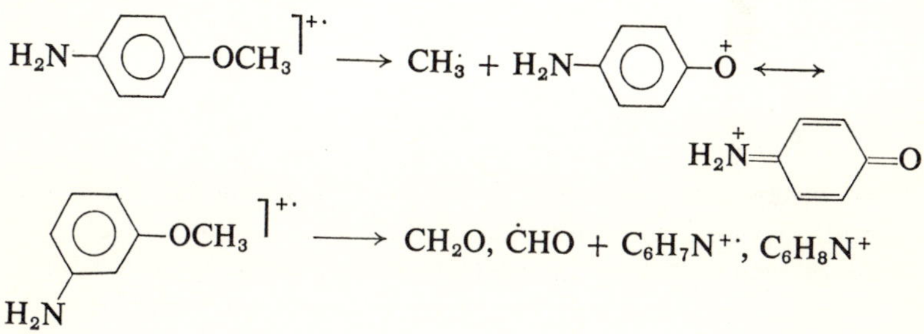

As in solution chemistry, *ortho*-substituents in benzene rings may behave differently from both the *meta*- and *para*-substituents. The difference in behaviour is often called an *ortho-effect*. The elimination of CH_4O from methyl anthranilate and of OH from *ortho*-nitrotoluene

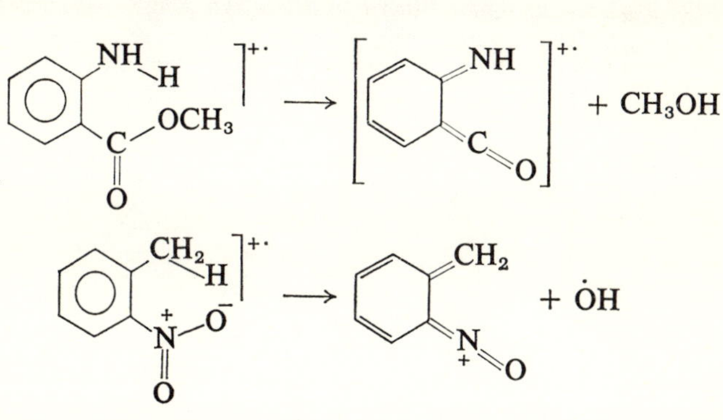

are examples of this *ortho*-effect. The process is quite general and may be represented by reaction scheme (4.24) in which A, B, C, D may be carbon, oxygen, nitrogen, and sulphur.

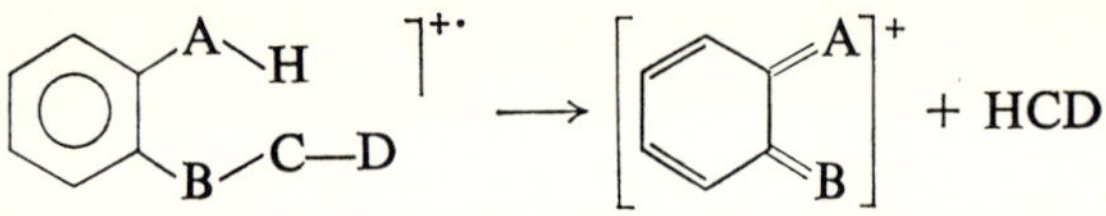

(4.24)

It is frequently observed that *ortho*-substituents are eliminated more readily than *meta*- or *para*-substituents and a cyclization process is often invoked to explain the process; but there is little evidence to support the suggestion. This ejection of *ortho*-substituents is also described as an *ortho*-effect.

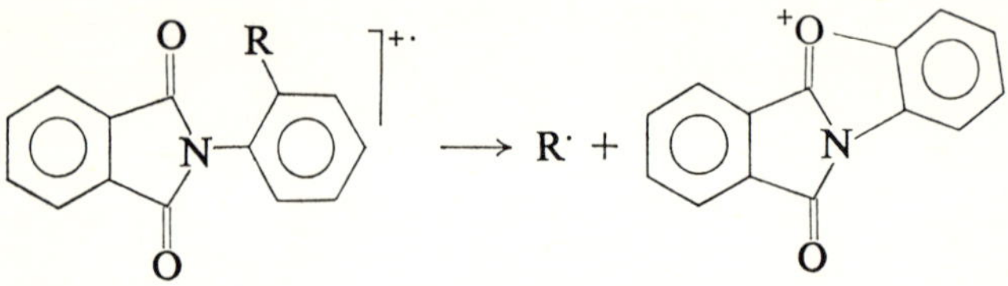

R = Cl, NO_2, . . .

For disubstituted benzenes the order of fragmentation at the substituents may be predicted quite accurately from table 4.8 which gives alongside each substituent the neutral particle commonly ejected during its fragmentation; generally, the higher a substituent in the table, the more readily fragmentation begins there. Thus, in a chloroacetophenone elimination of methyl is observed before chlorine and in a fluorobromobenzene loss of bromine is observed but not fluorine. The table is

TABLE 4.8 *Order of ease of fragmentation initiated by the presence of a substituent in a benzene ring*†

Substituent	Neutral particle eliminated
$COCH_3$	CH_3
CO_2CH_3	OCH_3
NO_2	NO_2
I	I
OCH_3 ⎤	CH_2O, CHO
Br ⎥	Br
OH ⎦	CO, CHO
CH_3	H
Cl	Cl
NH_2	HCN
CN	HCN
F	C_2H_2

† Decreasing ease of fragmentation from top to bottom of table. The bracketed substituents are closely similar in ease of fragmentation.

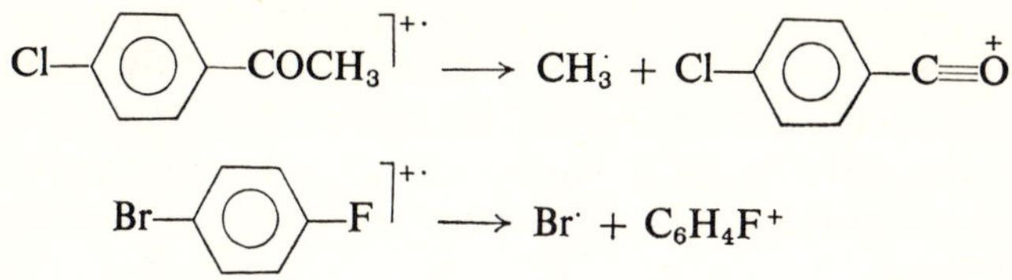

intended as a guide only because strong interaction effects between substituents can modify their behaviour as shown earlier. Benzenes substituted with OCH_3, Br, OH groups require very similar energies for fragmentation so that where these substituents occur together in a benzene compound, table 4.8 serves as only a nominal guide to the order of fragmentation.

The role of substituents in the mass spectra of aromatic compounds has been examined in some detail through the use of Hammett σ-constants. The elimination of C_2H_3O (CH_3CO?) from *meta-* and *para-*substituted acetophenones, $R \cdot C_6H_4COCH_3$, yields a fragment ion, $R \cdot C_6H_4^{+}$. In each mass spectrum of this series (R = NO_2, Cl, Br,

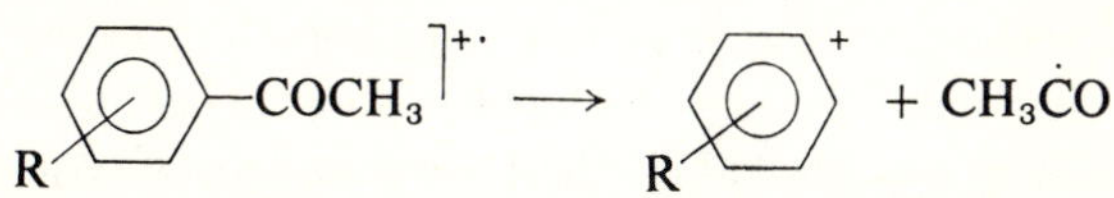

OCH_3, NH_2, etc.) the height (M) of the peak for the molecular ion was compared with the height (A) of the peak for the fragment ion. The ratio (A/M) was termed Z, and for R = H, the ratio Z becomes Z_0. When $\log(Z/Z_0)$ was plotted as ordinate with the corresponding Hammett σ-constants as abscissae, a good straight line correlation was observed. Similar correlations have been sought in other mass spectrometric fragmentations of aromatic compounds, but it is not clear what the ratio (Z/Z_0) represents. The heights of the peaks representing the molecular and fragment ions have been described as resulting from steady state conditions in the ion source i.e. the relative heights (A/M) are supposed to be a measure of the rate of fragmentation so that $\log(Z/Z_0)$ may be equated with $\log(k/k_0)$; k, k_0 are the rates of decomposition of the molecular ions in the series. In reaction kinetics, $\log(k/k_0)$ can be related by a linear free energy relationship to σ,

$$\log(k/k_0) = -\Delta\Delta G^{\ddagger}/RT = \rho . \sigma \qquad (\rho = \text{a constant})$$

It was assumed that if $\log(Z/Z_0) = \log(k/k_0) = \rho . \sigma$ then, $\log(Z/Z_0)$

could be a measure of the relative activation energies for the reactions. The heights of the peaks in a mass spectrum reflect mostly the number of ions with insufficient energy to decompose further but containing various amounts of excess of internal energy. If the ratio (A/M) is to be taken as a measure of the decomposition rate of the molecular ions (k) then it must be assumed to be a 'mean' rate since each ion in the mass spectrometer decomposes at its own rate, there being no equilibration of energy between ions (§3.3). There is doubt of the validity of assuming generally that $\log(k/k_0) = \log(Z/Z_0)$, and even where this relationship is true, it can be shown that observed correlations with σ may be due to nothing more than the known correlation of molecular ionization potentials with the Hammett σ-constants.

Lack of knowledge of ion structures and the electronic states leading to fragmentation and the uncertainty of what (Z/Z_0) actually represents provide good grounds for reluctance to draw conclusions from observed correlations with σ-constants. In attempts to improve the correlations, recourse to the use of Brown σ^+-constants in place of Hammett σ-constants has been made with some marginal improvement in some cases.

The appearance potential (§5.4) of a fragment ion in a mass spectrum represents the minimum energy required to produce just sufficient of the ions for them to be detected. If the fragmentation takes place in the ground state molecular ion, the difference between the appearance potential (AP) of the fragment ion and the ionization potential (IP) of the molecular ion represents the energy (E) required for fragmentation of the molecular ion. Figure 4.19 shows the relationship between these quantities for the reaction (4.25).

$$\text{R—M} \xrightarrow{-e} \text{R—M}^{+\cdot} \longrightarrow \text{R—A}^{+} + \text{B}^{\cdot} \qquad (4.25)$$

The reaction energy (E) has been plotted against both σ and σ^+ constants in various other fragmentations to give linear relationships in some instances. The method again suffers from lack of knowledge of ion structures and the electronic states leading to fragmentation. Additionally, there is some doubt of the accuracy of measurement of appearance potentials because these quantities may contain undefined excess of energy terms (§5.2). One example of the correlation of (AP – IP) with σ for the fragmentation (4.26) is illustrated in figure 4.20.

$$\text{R.C}_6\text{H}_4\text{NMeCOCF}_3]^{+\cdot} \longrightarrow \text{CF}_3\text{C}{\equiv}\overset{+}{\text{N}}\text{Me} + \text{R.C}_6\text{H}_4\dot{\text{O}} \qquad (4.26)$$

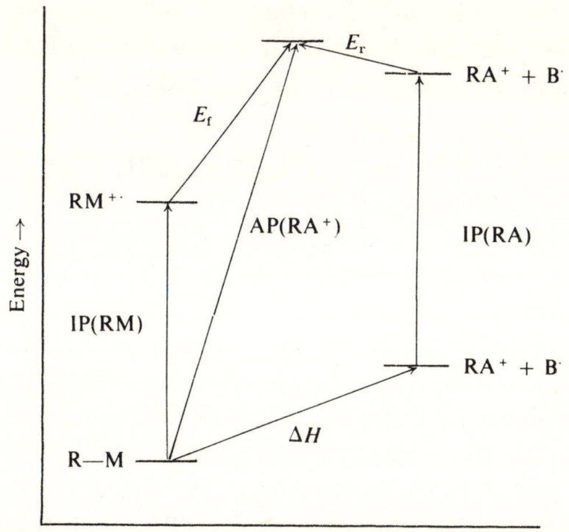

Figure 4.19. Diagrammatic representation of energy relationships for the simple bond cleavage, $RM^{+·} \rightarrow RA^+ + B^·$, occurring in the ground state with no excess energy terms. E_f, E_r = activation energies for forward and backward reactions; IP(RM), IP(RA) are ionization potentials for the molecule (RM) and the radical (RA); AP(RA^+) is the appearance potential for the ion (RA^+); ΔH is the bond dissociation energy for the reaction, $RM \rightarrow RA^· + B^·$.

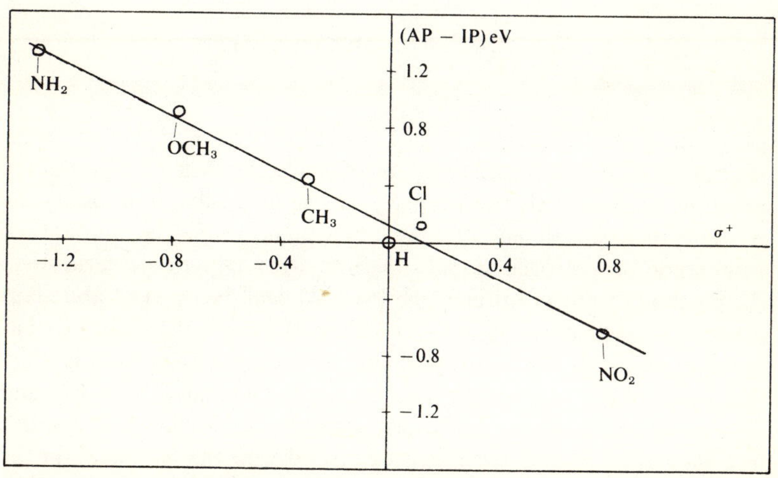

Figure 4.20. Correlation between (AP − IP) and σ^+ for the *para*-isomers in the reaction, $ArNCH_3 \cdot COCF_3^{+·} \rightarrow ArO^· + CH_3\overset{+}{N}{\equiv}CCF_3$.

(iii) Aromatic compounds like benzene yield relatively stable and abundant molecular ions. The most important initial fragmentation of benzene leads to the elimination of C_2H_2 and C_2H_3:

$$C_6H_6^{+\cdot} \nearrow C_2H_2 + C_4H_4^{+\cdot}$$
$$C_6H_6^{+\cdot} \searrow C_2H_3^{\cdot} + C_4H_3^{+}$$

Heteroaromatic compounds also give good molecular ions and, apart from side-chain or substituent fragmentations such as those described in (i) and (ii) above for benzene compounds their fragmentation leads to the ejection of the heteroatom in a neutral species. Whereas benzene ejects C_2H_2 and C_2H_3 from the molecular ion, pyrrole and pyridine eject HCN. Similarly, thiophene ejects CHS and furan eliminates CHO from their molecular ions. Fragmentations of the heteroaromatic

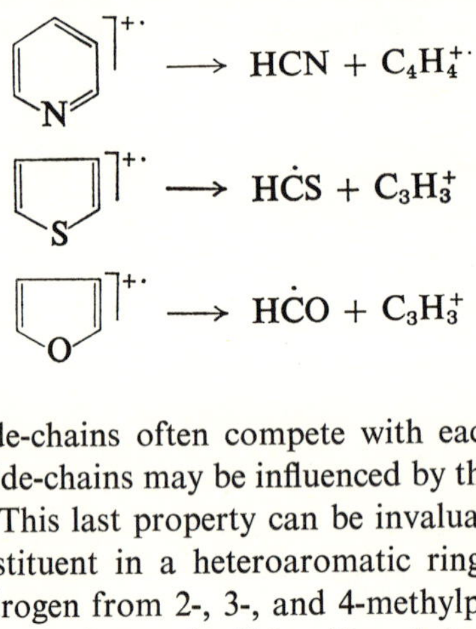

ring and of its side-chains often compete with each other; the fragmentation of the side-chains may be influenced by their positions in the aromatic nucleus. This last property can be invaluable in deciding the position of a substituent in a heteroaromatic ring. The amount of elimination of hydrogen from 2-, 3-, and 4-methylpyridine varies with the position of the substituent the [M – 1] peak being most abundant in the 3-methyl isomer (figure 4.21). Similarly, the 2-methyl isomer has the most abundant [M – 15] peak corresponding to the loss of methyl. The pyridine-2,3-dicarboxylic acid imide (1) eliminated both CO and CO_2 from the molecular ion to give ions at *m/e* 134, 118 with relative abundances of 17 and 4 per cent compared with the base peak in the spectrum (reaction 4.27). The 3,4-isomer (2) afforded ions at the same *m/e* values but with relative abundances of 1 and 27 per cent.

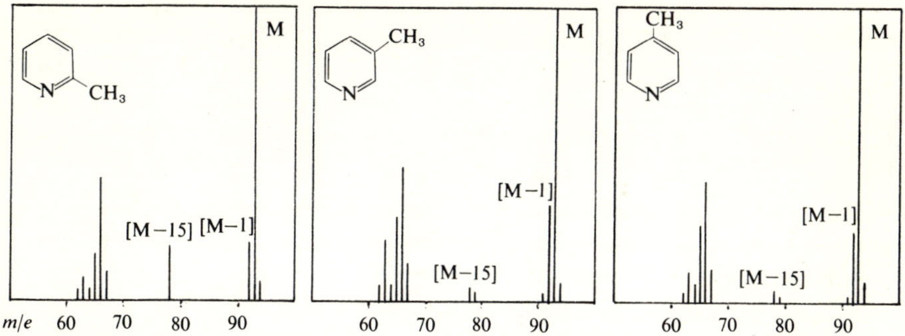

Figure 4.21. Mass spectra of 2-, 3-, and 4-methylpyridine showing enhanced abundance of [M-1] peak in the 3-methyl isomer, and enhanced abundance of [M-15] peak in the 2-methyl isomer.

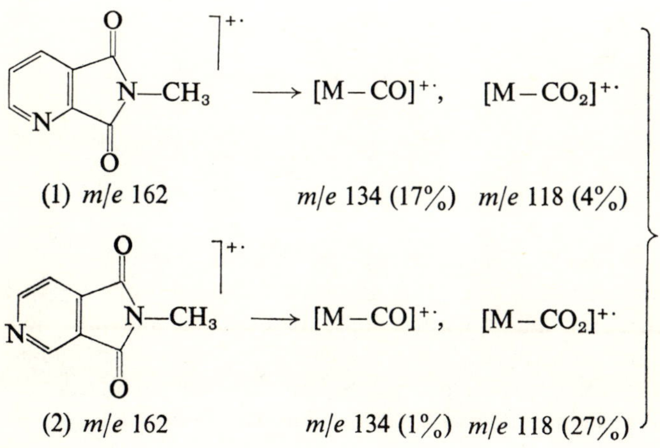

(4.27)

Deuterium labelling strongly suggests that extensive rearrangement of benzene compounds occurs in the molecular ion before fragmentation. It has been suggested that these rearrangements proceed through benzvalene type intermediates. Sometimes this sort of rearrangement is apparent from the nature of the fragment ions. For example, the mass spectra of oxazoles contain features very similar to those of isoxazoles and it has been suggested that rearrangement occurs via an azirane (4.28) as found in the photochemistry of these compounds.

Because C2 type side-chain cleavage of heteroaromatic compounds is often prominent in their fragmentation, it has been suggested that the fragment ions are analogous to the supposed tropylium structure found

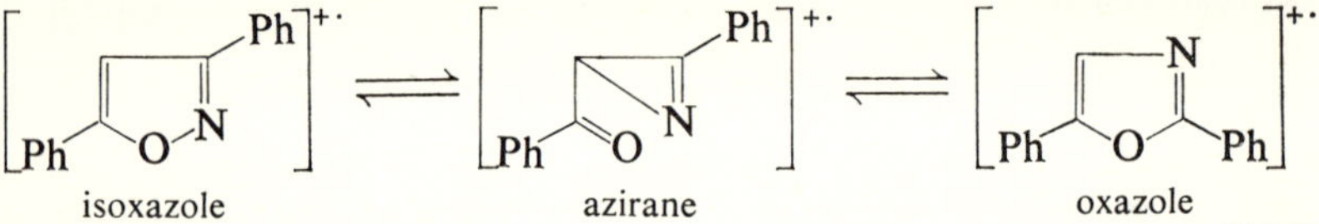

isoxazole azirane oxazole

(4.28)

with benzyl compounds i.e. that ring expansion occurred. However, there are no firm experimental details to support this contention and there is even doubt on the correctness of the tropylium structure for the $C_7H_7^+$ ions from benzylic compounds. Nevertheless, in this type of fragmentation behaviour, heteroaromatic compounds are very similar to simple benzene compounds.

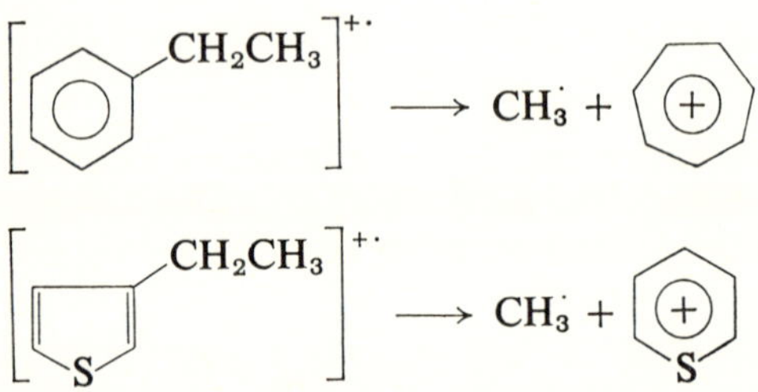

The organometallic 'sandwich' compounds like ferrocene exhibit many characteristics of the fragmentation behaviour of aromatic compounds. Abundant molecular ions are observed and any side-chains present normally fragment like their simple benzene counterparts. Also, just as heteroaromatics decompose by fragmentation of the aromatic ring, the metallocene molecular ions decompose with destruction of the sandwich structure to leave an ionized metal (M).

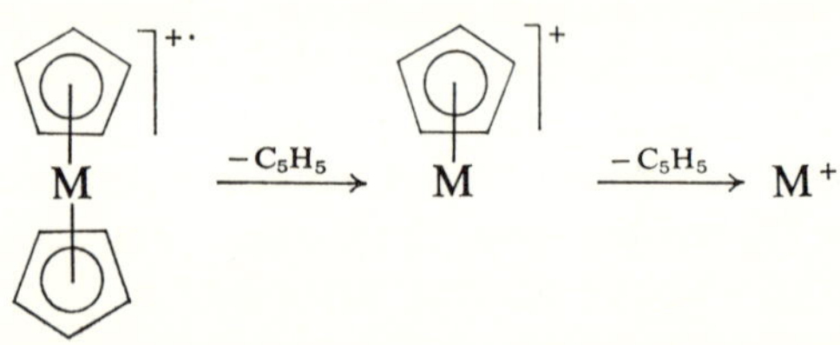

In summary, aromatic compounds exhibit many of the features found in the mass spectra of aliphatic compounds but are usually considerably modified because of (*a*) the increased stability of the molecular ions, (*b*) the aromatic ring system itself behaving as a substituent (X) in an alkane, (*c*) substituents on the aromatic ring changing fragmentation behaviour, and (*d*) heteroatoms in the aromatic ring influencing the side-chain fragmentation depending on its position.

4.8. Subsequent decomposition of primary fragment ions. Most molecular ions formed by electron-impact are radical–cations having one unpaired electron. Radical–cations are known in solution chemistry to be reactive species and the molecular ions formed in mass spectrometry frequently decompose by ejection of a radical to form an even-electron charged species. In even-electron species, the electrons can pair in orbitals to give closed shells; such even-electron cations in solution

$$[CH_3CH_2CH_3]^{+\cdot} \longrightarrow CH_3^{\cdot} + C_2H_5^{+}$$

$[CH_3CH_2CH_3]^{+\cdot}$	$CH_3^{\cdot}$	$C_2H_5^{+}$
m/e 44 (even mass number)		*m/e* 29 (odd mass number)
Radical–cation (odd number of electrons)	Radical	Cation (even number of electrons)

chemistry, the carbonium ions, are generally less highly reactive than odd-electron species like radical–cations.

If the radical–cation does not eject a radical to produce an even-electron ion, rearrangement may occur with subsequent elimination of a non-radical fragment to give another odd-electron ion. The rearranged odd-electron fragment can then decompose by ejection of a radical to

$$[C_6H_5OCO \cdot OCH_3]^{+\cdot} \longrightarrow CO_2 + C_7H_8O^{+\cdot}$$

m/e 152 (even mass number)
Radical–cation
(odd number
of electrons)

$C_7H_8O^{+\cdot}$
m/e 108 (even mass number)
Radical–cation

$$\downarrow -\dot{C}HO$$

$C_6H_7^{+}$
m/e 79 (odd mass number)
Cation
(even number
of electrons)

yield an even-electron ion as in the example of methyl phenyl carbonate shown here. It is found that many of the major ions in a mass spectrum occur at odd *m/e* values (even-electron ions). Ions at even *m/e* values (odd-electron species) are produced by rearrangement (for odd numbers of nitrogen atoms in an ion this behaviour is reversed i.e. odd *m/e* values correspond to odd-electron species and vice versa).

It will be appreciated from the foregoing that subsequent decomposition of primary fragment ions usually revolves around the behaviour of even-electron species. It may be recalled that one of the first items to be noted in a mass spectrum is which of the major fragment ions occur at odd *m/e* values and which at even *m/e* values.

The initial fragmentation reactions of molecular ions can be grouped into a few general schemes but the decomposition reactions of the fragments themselves are not so easy to classify. The first fragmentations of a molecule in a mass spectrometer are generally those giving most structural information because after several decompositions along any particular fragmentation pathway, the relationship between the ions and the original structure becomes too tenuous to be of much value. Fragment ions (A^+, B^+) from a molecular ion (M^+ in 4.29) may

$$M^+ \longrightarrow A^+ \longrightarrow B^+ \longrightarrow C^+ \longrightarrow D^+ \longrightarrow \qquad (4.29)$$

yield valuable information about M^+ e.g. loss of H_2O, CH_3, CH_3O, etc., but by the time fragmentation reaches the ion D^+ much of the original structure of M^+ has been destroyed and little structural information may be obtained. However, ions at the low mass end of a spectrum should not be ignored as important clues to the structure of M^+ may still be obtained e.g. $C_2H_3O^+$ from CH_3CO groups at *m/e* 43, CH_4N^+ at *m/e* 30 from primary amines. Often such useful low mass ions arise from primary fragmentation of the molecular ion. The loss of C_2H_3O from a molecular ion can be recognized by the presence of an $[M-C_2H_3O]^+$ ion as well as the $C_2H_3O^+$ ion at *m/e* 43. The later stages of a fragmentation pathway frequently only become clear after the structure of a compound has been elucidated. These later fragment ions then serve as confirmation for a proposed structure; compare the use of the 'finger-print' region in infrared spectroscopy which may be used for confirmatory evidence for an assignment but is not very useful for initial diagnosis. Aliphatic fragment ions decompose further by elimination of C_2H_4.

$$R\cdot CH_2\cdot CH_2^+ \longrightarrow R^+ + C_2H_4$$

Heteroatom molecules, after initial fragmentation, often decompose further after hydrogen transfer. The migrating hydrogen may arise from a β-, γ-, or δ-carbon from the heteroatom. The common C2 cleavage reaction of heteroatomic compounds yields an even-electron fragment ion which, after the hydrogen transfer, may fragment along two different pathways to give even-electron ions containing the heteroatom. The processes may be represented generally by scheme (4.30) in which X is nitrogen, oxygen, or sulphur. Ions at *m/e* 30, 58 in the mass

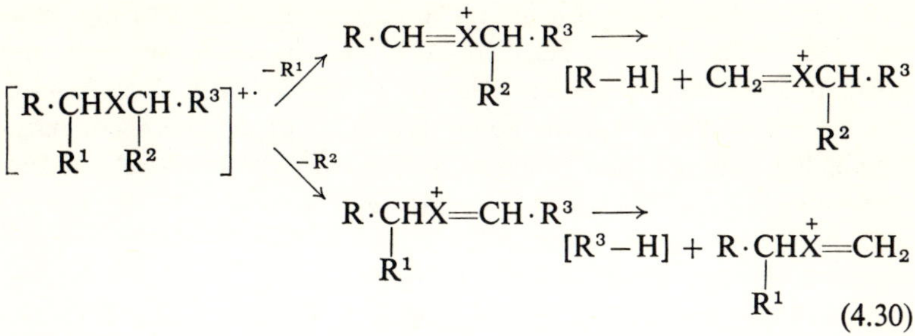

(4.30)

spectrum of ethyl-n-propylamine (4.31) are examples of this type of fragmentation.

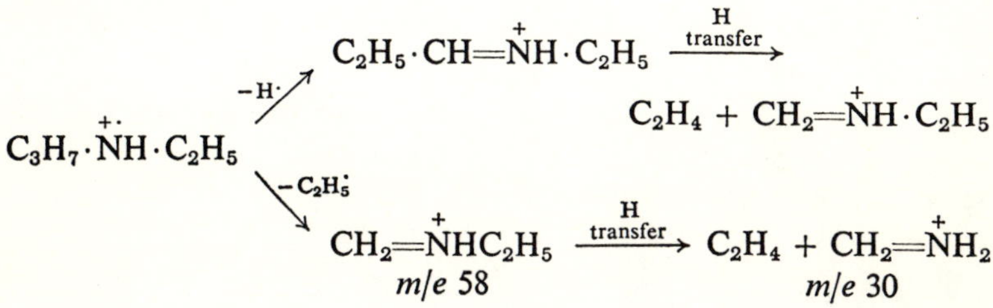

(4.31)

Hydrogen transfer may also occur with elimination of the heteroatom as the neutral species and is quite common for halides and esters. For example, if X = Br (—CH(R^2)R^3 absent in scheme 4.30) then HBr is eliminated as in reaction sequence (4.32).

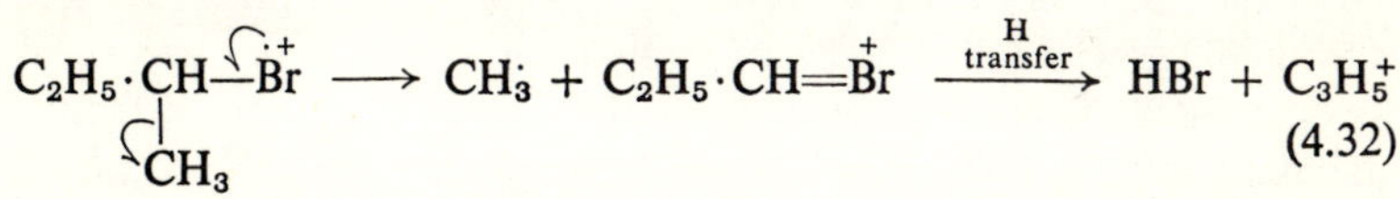

(4.32)

The elimination of a neutral fragment from even-electron ions is particularly marked in the mass spectra of many simple organometallic compounds. Ejection of a C_4H_9 radical from the molecular ion of tetra-n-butyl tin affords an even-electron ion which then eliminates C_4H_8 with hydrogen transfer to give a second even-electron ion; this latter ion then decomposes again by elimination of a second C_4H_8 unit to yield a third even-electron ion.

$$(C_4H_9)_4Sn^{+\cdot} \xrightarrow[-C_4H_9]{} (C_4H_9)_3Sn^{+} \xrightarrow[-C_4H_8]{\text{H transfer}} (C_4H_9)_2SnH^{+} \xrightarrow[-C_4H_8]{\text{H transfer}} (C_4H_9)SnH_2^{+}$$

When rearrangement of a molecular ion accompanies its fragmentation, the product ion may itself fragment like the molecular ion of the postulated rearranged species. The point may be clarified by two examples. Ejection of CO from the molecular ion of the pyranocoumarin (1) yields a fragment ion which is conveniently postulated as the furanocoumarin ion (2). The subsequent decompositions of the ion (2) and of the molecular ion produced by a specimen of the furanocoumarin itself were almost identical. Analogously, the methoxyphenazine (3) eliminated CH_2O from its molecular ion to give a fragment ion postulated as the phenazine molecular ion (4). The subsequent

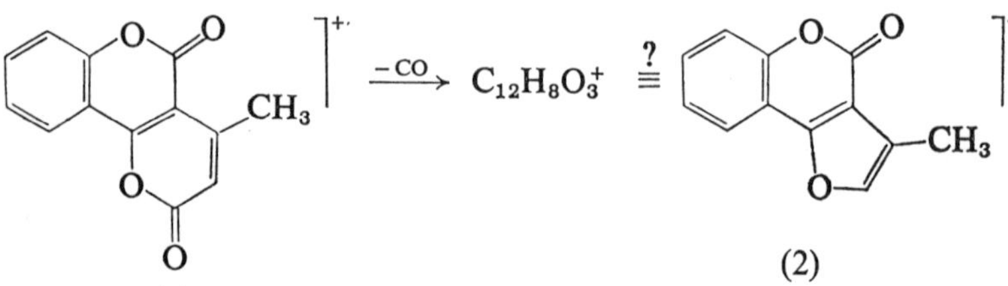

(1) (2)

Subsequent decompositions are very similar

decomposition of the ion (4) was almost identical to that of the molecular ion of phenazine. The close similarity in behaviour of the respective fragment and molecular ions suggests they *may* have similar structures although, especially for the decomposing ion, these are not necessarily as postulated. There is thermochemical evidence that some at least of the many observed similarities of ion decompositions are accidental and that the ion structures cannot be as postulated.

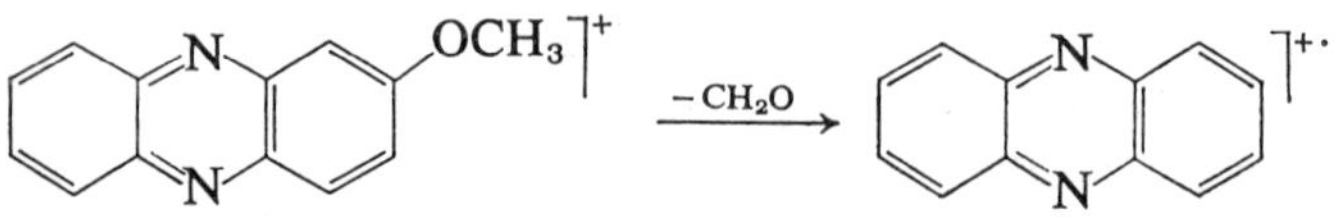

(3) (4)

Subsequent decompositions are similar to those of molecular ion of phenazine

4.9. Rearrangement accompanying fragmentation. The formation of new bonds between atoms in an ion leads to rearrangement. Hydrogen migration, which seems particularly easy, has been covered in earlier sections. In some instances, mainly with odd-electron ions, the site from which a hydrogen is transferred is fairly specific. In other cases, generally with even-electron ions, the reaction appears to be far less specific. The randomization of hydrogen in ions is widespread and can make the interpretation of deuterium labelling studies difficult or ambiguous.

Skeletal rearrangement is defined as the formation of new bonds between atoms other than hydrogen.

Skeletal rearrangement. The classification of these rearrangements is useful for the interpretation of mass spectra because known structural features giving rise to this phenomenon can be compared with the behaviour of compounds of unknown but postulated structure. Skeletal rearrangements in mass spectrometry are diverse but attempts have been made to classify them.

One of the earliest and most useful classifications was based on (i) the migration of alkyl or aryl groups, and (ii) the nature of the neutral particle ejected after rearrangement (Brown and Djerassi, 1967). These compilations are useful, and for structural organic chemistry the nature of the ejected species may be particularly informative. Typical neutral species eliminated following rearrangement include CO, CO_2, SO_2, CH_2O, HCN, S and all have favourable negative heats of formation. The diagnostic value of the ejected neutral particle may or

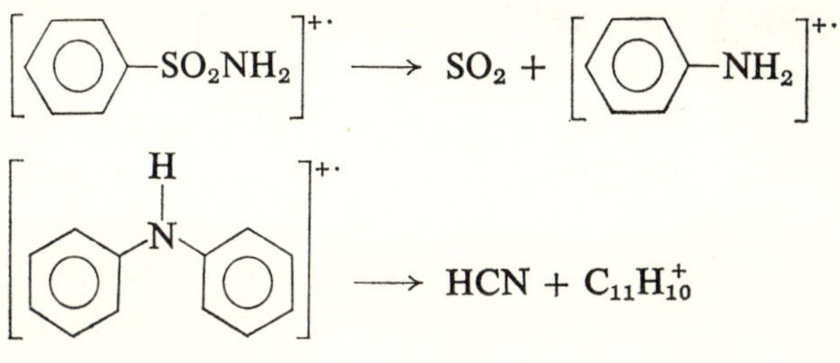

may not be significant. Thus, elimination of CO_2 from a molecular ion indicates an unsaturated ester, carbonate, or cyclic imide group as likely structural features but loss of CO occurs from a wide variety of compounds and yields less immediate information. The elimination of CO_2 from unsaturated esters and carbonates is quite general, although the degree to which it is observed is very variable. The process has been

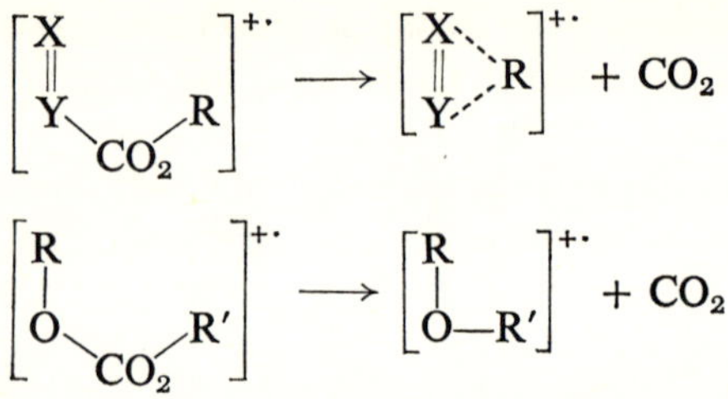

represented by two schemes in which X, Y are doubly or triply bonded in the unsaturated esters and either R or R′ must be aryl in carbonates.

It has been suggested that rearrangement processes may be grouped into the reaction scheme (4.33) in which A, C are usually unsaturated

$$[\mathrm{A{-}B{-}C}]^+ \begin{cases} \nearrow [\mathrm{A{-}C}]^+ + \mathrm{B} \\ \searrow [\mathrm{A{-}C}] + \mathrm{B}^+ \end{cases} \qquad (4.33)$$

and B ejected as a neutral particle. The scheme is too general to be useful but the possibility that an electron-deficient site on A can provide a place to which C could migrate might be more useful.

The idea of migration to a positive centre *via* a nominal four-centre process in even-electron ions is embodied in scheme (4.34) two examples

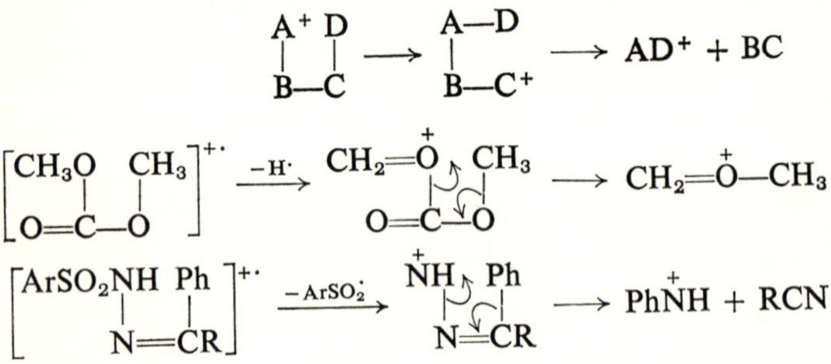

(4.34)

of which are shown. Whereas these rearrangements appear to be four-centre ones, the corresponding reaction of trimethylsilyl ethers seems amazingly insensitive to distance between the reacting centres.

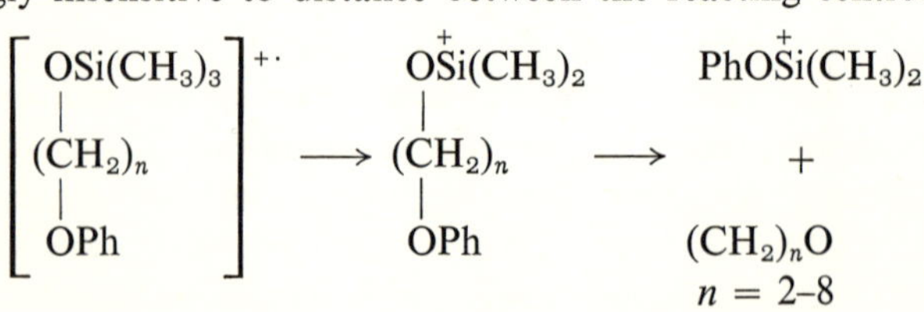

A variety of formally four-centre rearrangements in odd-electron ions can be collected in one scheme (4.35), very similar to that of (4.34) but migration to a positively charged centre is not implied. The main requirement for this reaction appears to be that atoms A, B, C, D

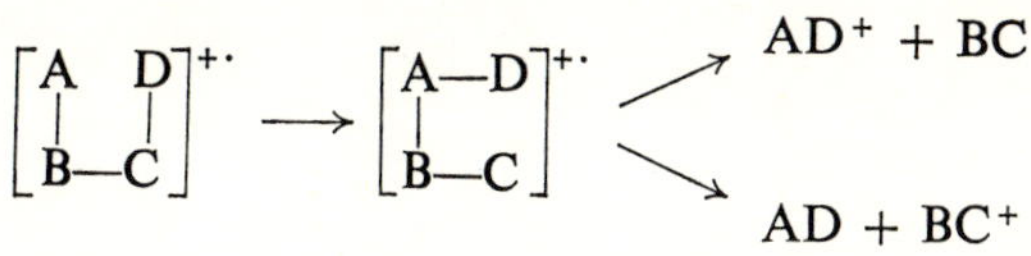

(4.35)

which may be nitrogen, oxygen, sulphur, or carbon should form contiguous π-centres. It is not suggested that all these rearrangements actually proceed through a four-membered transition state but when four such π-centres are present in a molecule, then rearrangement ions may be expected in the mass spectrum. Two examples from the mass spectra of trifluoroacetyl amides and cyclic thioimides illustrate the general nature of this rearrangement.

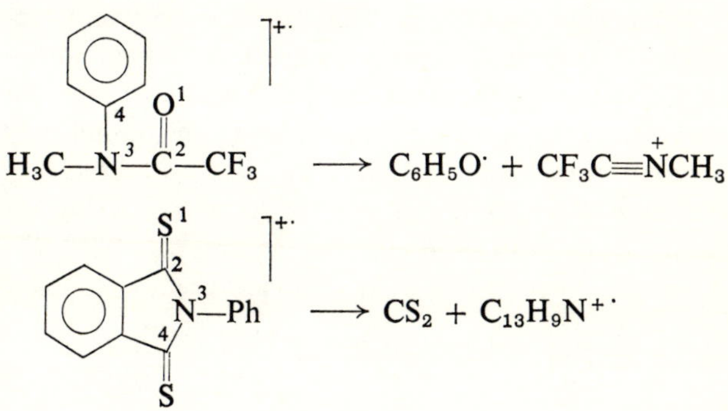

Often rearrangement competes with simple cleavage and the latter may be so easy that peaks representing rearrangement processes are of minor intensity or not present at all. From this point of view, the closely similar amides (1, 2) in scheme (4.36) are interesting since the first (1) shows almost no rearrangement whereas the second (2) has abundant ions corresponding to substantial rearrangement. Unlike the nominal four-centre rearrangement of even-electron ions, migration of alkyl groups seems uncommon in odd-electron ions.

It was pointed out many years ago (Field and Franklin, 1957) that rearrangement processes would have lower frequency factors (see §3.4) than simple bond breaking reactions. Subsequent experiments have

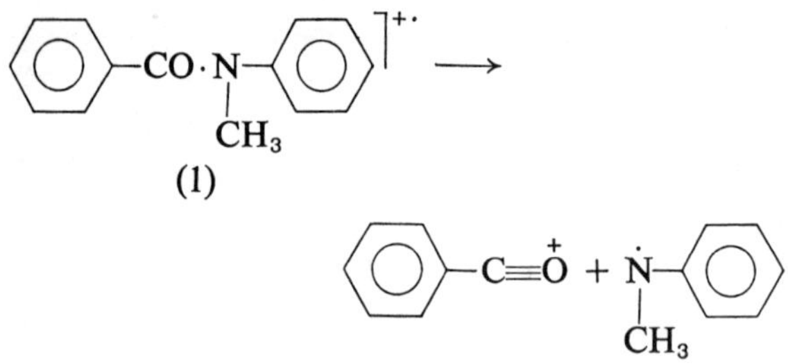

Simple cleavage

(4.36)

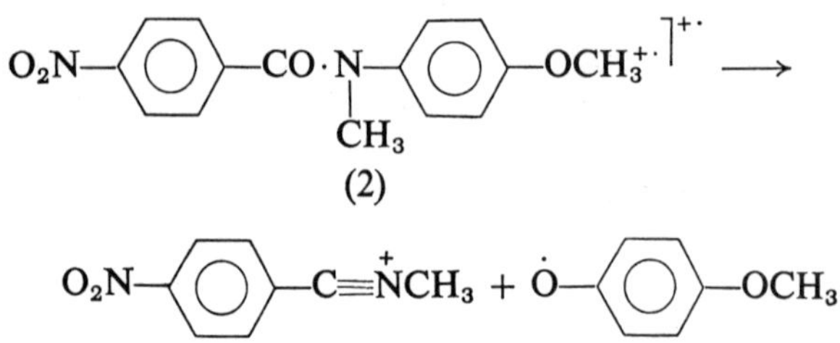

Rearrangement

demonstrated the correctness of this statement. Thus, by gradually reducing the electron-beam energy in electron-impact ionization of molecules, it is observed that rearrangement reactions may increase in importance relative to simple cleavage. Using the simplified form of the quasi-equilibrium rate equation, the rate of a rearrangement–cleavage reaction (k_R) may be compared with the rate of simple cleavage (k_C) for a molecule with N oscillators. In the equations (4.37), the frequency factors for rearrangement with cleavage and simple cleavage

$$k_R = \nu_R\left[1 - \frac{E_R^0}{E}\right]^{N-1}; \qquad k_C = \nu_C\left[1 - \frac{E_C^0}{E}\right]^{N-1} \tag{4.37}$$

are ν_R, ν_C respectively and the energies of activation for the processes E_R^0, E_C^0. The internal energy in the ion is E. Equations (4.37) combined give (4.38) which, as a first approximation, may be written as in equation (4.39) where A is a constant and $\Delta E = E_C^0 - E_R^0$. When the

$$\log(k_R/k_C) = \log(\nu_R/\nu_C) + (N - 1)\log\left(1 - \frac{E_R^0}{E}\right)\Big/\left(1 - \frac{E_C^0}{E}\right) \tag{4.38}$$

$$\log(k_R/k_C) \approx \mathrm{A} + (N - 1)(\Delta E/E) \tag{4.39}$$

energy of activation for simple cleavage is greater than the energy of activation for rearrangement with cleavage, then $\Delta E/E$ in (4.39) is positive and $\log(k_R/k_C)$ increases as E is reduced; i.e. reducing the electron-beam energy, and therefore the energy transferred to the molecular ion causes the ratio of the rate of the rearrangement process relative to the rate of the simple bond breaking reaction to increase. Therefore, as the electron-beam energy decreases the importance of the rearrangement process increases relative to that of simple bond fission. The same conclusion may be reached by comparison of the rates k_R, k_C at different internal energies of an ion (figure 4.22). Using the

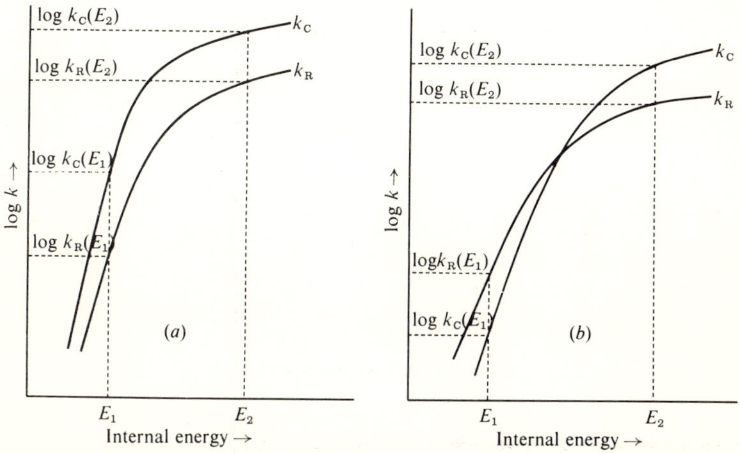

Figure 4.22. Variation of rates of simple bond cleavage (k_C) and of rearrangement with cleavage (k_R) with changes in internal energy of an ion. In (*a*) the curves do not cross; in (*b*) the curves do cross.

simplified rate equations (4.37) graphs of log k_R and log k_C plotted for various internal energies appear as in figure 4.22*a*. In this case, at all internal energies (e.g. E_1, E_2) log k_C is greater than log k_R and the rate of simple bond fission always exceeds that for rearrangement followed by fission. With energies of activation and frequency factors different from those obtaining in figure 4.22*a* the curves for log k_R and log k_C may cross (figure 4.22*b*). Then, at some low excess of internal energy (E_1), the rate of rearrangement with cleavage is faster than the rate of simple bond cleavage but at higher internal energy (E_2) the order of rates is reversed. Thus, at low electron-impact energies rearrangement is favoured over ordinary bond cleavage and, by reducing the electron-impact energy below the normal 70 volts, the increased importance of the rearrangement process may be observed by an increased abundance of ions corresponding to it.

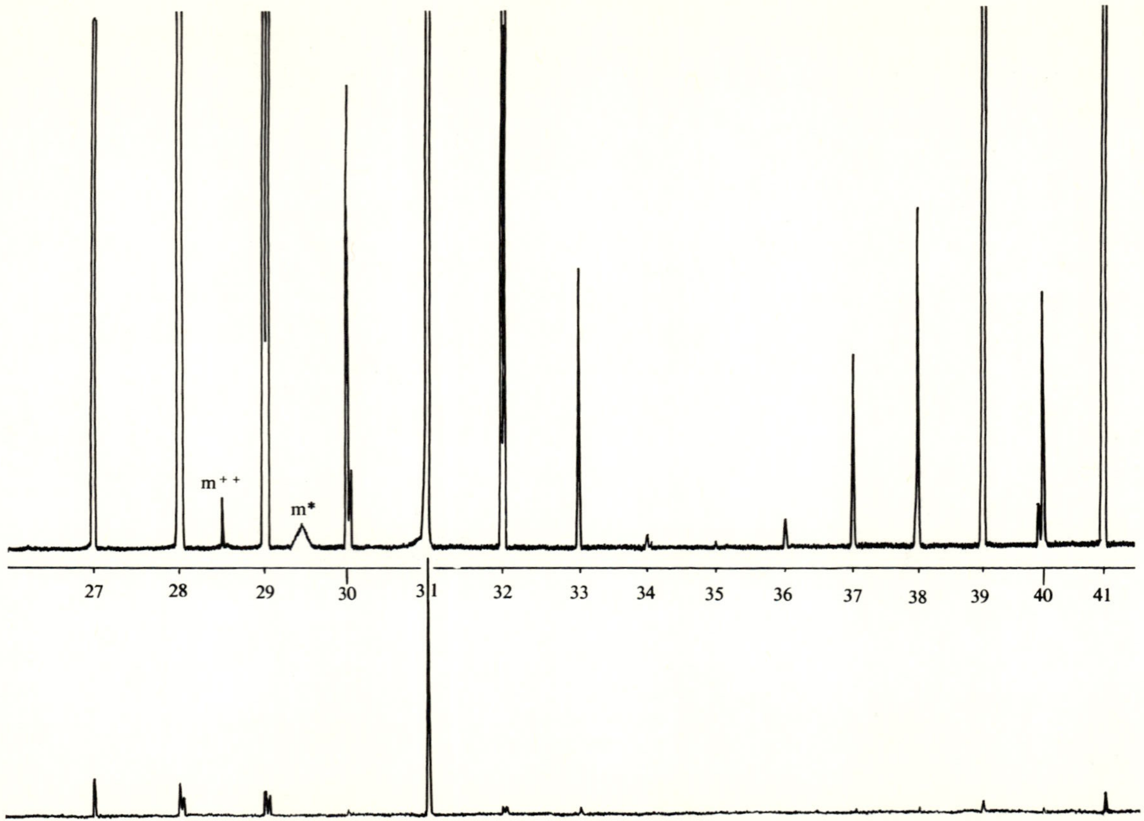
m⁺⁺
m*
27
28
29
30
31
32
33
34
35
36
37
38
39
40
41

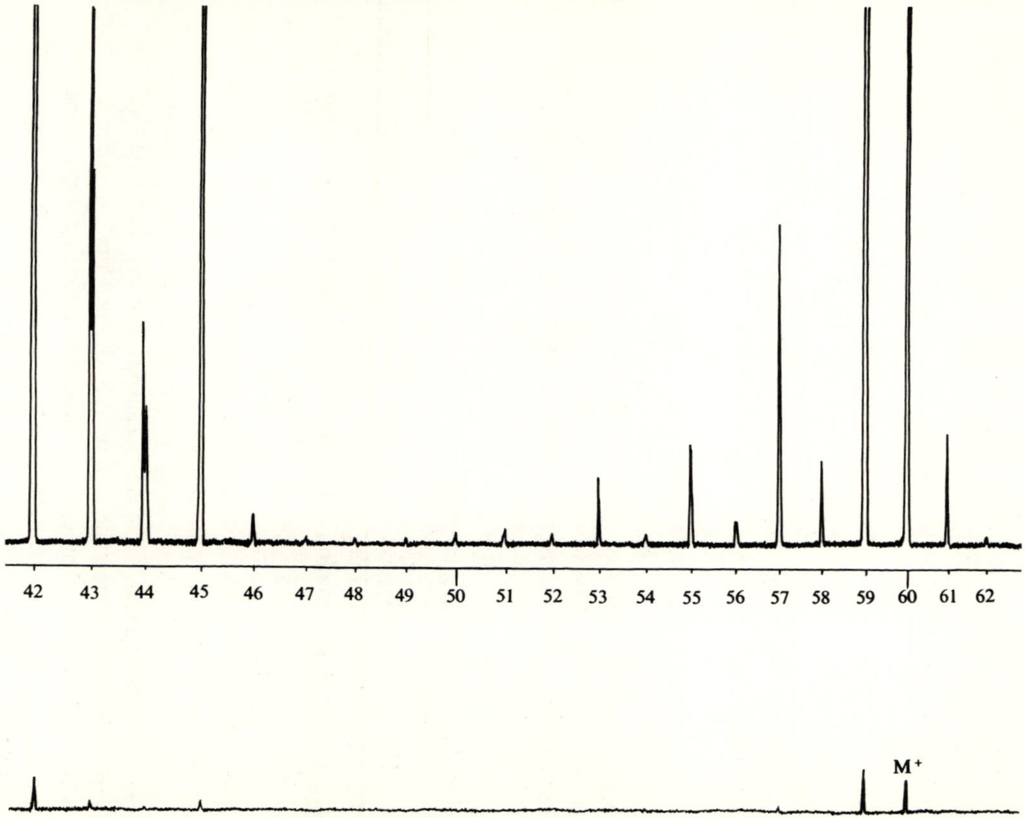

Figure 4.23. Mass spectrum of compound A.

Ions formed by field-ionization are collected very soon after formation and it has been suggested that there is too little time for much rearrangement to occur. It was expected that comparison of field- and electron-impact-ionization mass spectra would show the presence of rearrangement processes only in the latter. However, there is evidence that rearrangement, sometimes substantial, can occur in field-ionization spectra (see §5.10).

4.10. Examples of structure elucidation by mass spectrometry. A number of mass spectra are given below as normalized line diagrams together with any relevant accurate masses of ions. The first three examples are simple and are used to illustrate the approach to structure determination. The first example is worked through in some detail to show how the initial spectrum is examined. The remaining examples have not been chosen so that a complete structure could be deduced solely from the mass spectrum since this is infrequently the case in practice except for simple molecules or for 'families' of compounds the structures of some of which are known. Rather, the examples have been chosen to illustrate how mass spectrometry provides structural information for use in conjunction with other physical methods or chemical evidence of structure. In each instance the amount of information that can be easily inferred from the mass spectrum is indicated.

A. The actual mass spectrum of compound A is reproduced in figure 4.23; only two traces are given for clarity and simplicity. The upper trace is used for counting the spectrum and finding metastable ion peaks and the lower trace is used for structural work. The upper trace was obtained at a sensitivity 100 times greater than for the lower trace. The peaks at *m/e* 28,32 were identified by comparison with another previously calibrated spectrum. The spectrum is first counted on the upper trace. Notice how the gap between any two peaks slowly decreases as the *m/e* values increase; a sharp change between two successive gaps indicates the presence of a doubly charged ion e.g. see *m/e* 28.5 which is the doubly charged ion of mass 59. Counting stops at the last major ion, *m/e* 61, which must be the M + 1 ion (^{13}C isotope) and therefore the molecular ion is at *m/e* 60. There is a broad metastable ion peak at *m/e* 29.4. On the lower trace, many peaks of less abundant ions are not recorded and without the upper trace it would be difficult to count the spectrum accurately.

For convenience here, the mass spectrum is converted now into a normalized line diagram, although usually this would only be done for record purposes after the spectrum had been interpreted. The peak heights on the lower trace are measured (column 2, table 4.9). Because

TABLE 4.9 *Measurements taken from the mass spectrum of compound* A (*figure 4.23*)

m/e	Peak height (mm)	Relative abundance
60	6	13
59	8	17
45	2	4
43	1.5	3
42	6	13
41	4	8
40	0.5	1
39	2	4
33	1.5	3
32	1.5	3
31	47	100
29	5.5	11
28	6.5	14
27	7	15

traces intermediate between highest and lowest sensitivity have been omitted from the mass spectrum for the sake of simplicity, some of the peak heights measured on the lower trace cannot be checked (§1.6) and smaller peak heights are only approximate. The biggest peak (base peak) in the spectrum is at *m/e* 31 and this is put equal to 100 units (column 3, table 4.9); all other peaks are related to this base peak to give their relative abundances and the normalized line diagram (figure 4.24)

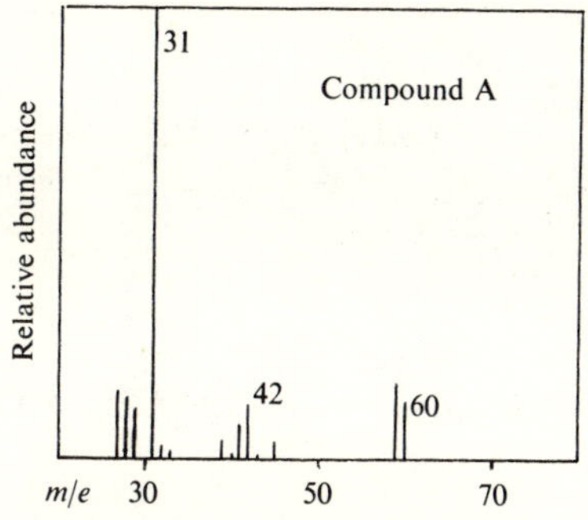

Figure 4.24. Normalized line diagram taken from mass spectrum of compound A in figure 4.23.

is then drawn. Peaks of less than 1 per cent of the abundance of the base peak have been omitted (e.g. *m/e* 44, see §1.1) and also peaks below *m/e* 27.

Amongst the major ions are *m/e* 59 (strong loss of one mass unit, H, from the molecular ion), *m/e* 45 (ejection of 15 mass units, CH_3), and *m/e* 31 (possible 'characteristic' ion of composition CH_3O?– §4.1). The compound A is of even molecular weight and therefore cannot contain an odd number of nitrogen atoms and the more abundant fragment ions are at odd *m/e* values (even-electron ions). One significant peak occurs at even mass, *m/e* 42, and is an odd-electron ion suggesting that rearrangement has occurred with loss of 18 mass units (H_2O) from the molecular ion. This last piece of information indicates that compound A is an alcohol and the loss of H from the molecular ion points to at least one hydrogen on the α-carbon atom (§4.6; C2 cleavage). The loss of H_2O also indicates a hydrogen on a

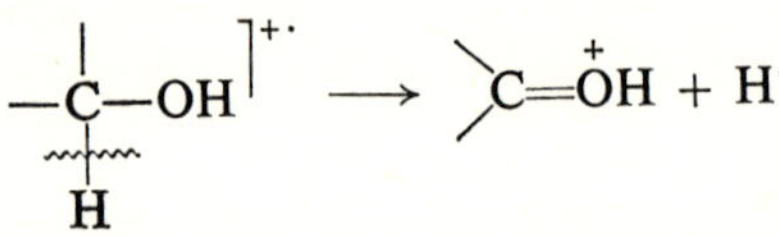

γ- or δ-carbon atom (§4.6; RE1 reaction). Since a δ-carbon atom implies four carbon atoms of 4 × 12 = 48 mass units, there would be insufficient mass remaining to accommodate an oxygen atom in the molecule. Therefore, there must be only three carbon atoms and a γ-hydrogen is ejected with the hydroxyl on fragmentation; the remaining hydrogens must fit as shown and compound A is n-propanol. The ion at *m/e* 31 confirms the primary alcohol character of A.

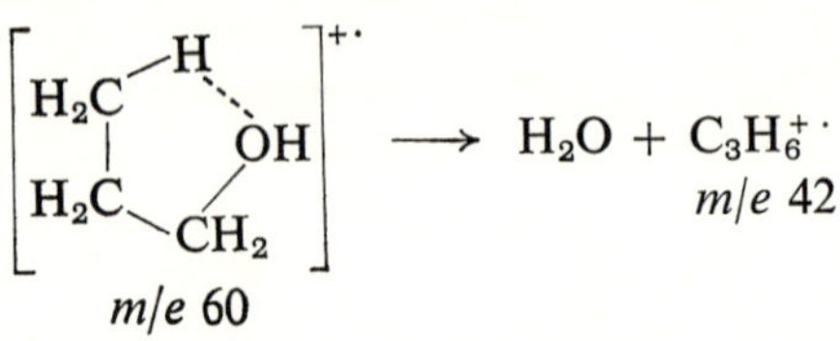

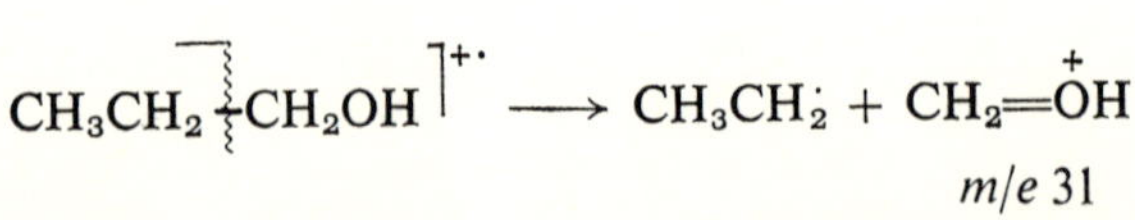

For comparison, it may be noted that isopropanol, which does not have a γ-carbon atom, ejects OH and not H_2O from the molecular ion.

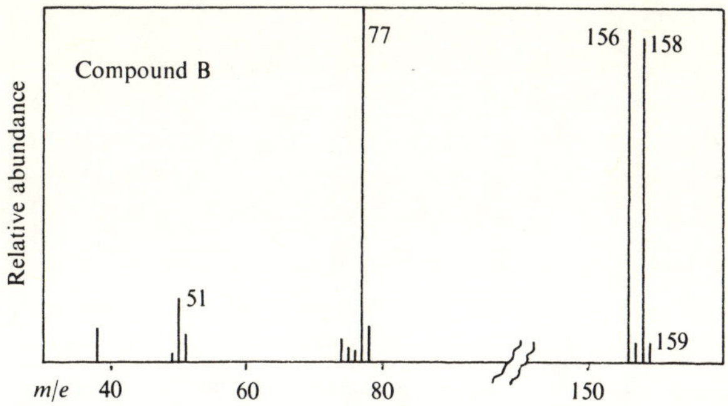

Figure 4.25. Mass spectrum of compound B. Ratio of peaks *m/e* 159:*m/e* 158 = 6.2:92.5.

B. Compound B gives two molecular ions at *m/e* 156, 158 almost of equal intensity (figure 4.25) and bromine must be present (the isotopes ^{79}Br and ^{81}Br are of almost equal natural abundance). Taking 79 and 81 mass units from 156 and 158 respectively leaves in each case 77. An abundant ion is observed at *m/e* 77; this is often a 'characteristic' ion of aromatic systems and, taken together with the general appearance of the mass spectrum (few but abundant fragment ions present), indicates that the compound is an aromatic bromide. The ratio of *m/e* 159 to *m/e* 158 is 6.2:92.5 or, stated in another way, the ^{13}C isotope peak at *m/e* 159 is 6.7 per cent of the height of the peak at *m/e* 158. Since each carbon atom contributes approximately 1.08 per cent of ^{13}C isotope, simple division (6.7/1.08 = 6, to nearest integer) gives the number of carbon atoms in compound B as six (§1.3). Thus, the unknown must have the composition, C_6H_5Br, and is probably bromobenzene. Compound B *is* actually bromobenzene but from the mass spectrum alone it is only possible to suggest this identification as other isomers, such 'Dewar' bromobenzene, for example, could not be ruled out on the evidence of the mass spectrum alone.

C. The mass spectrum of compound C shows a molecular ion at *m/e* 100 and accurate mass measurement gives 100.0889. Reference to tables or calculation (see table 1.4) affords the molecular composition, $C_6H_{12}O$(C_6 = 72.0000; H_{12} = 12.0938; O = 15.9949; total = 100.0887; error in measurement = 2 parts in 1 000 887 i.e. <2 p.p.m., which is acceptable). The composition, $C_6H_{12}O$, has one double bond equivalent (§4.2) which is accounted for by the carbonyl absorption in the infrared spectrum. Compound C must be a straight-chain aldehyde or ketone.

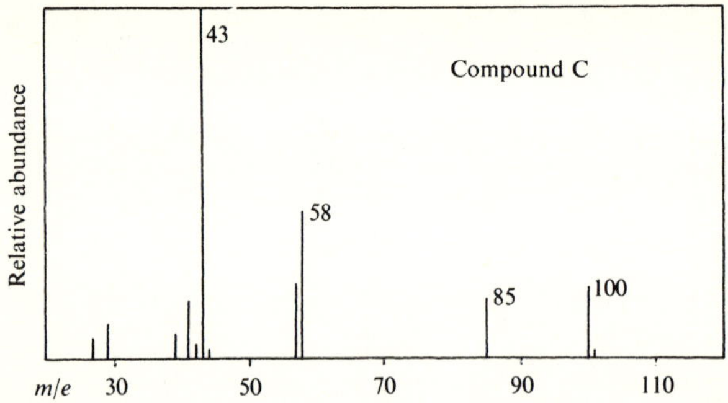

Figure 4.26. Mass spectrum of compound C. Accurate mass of molecular ion at *m/e* 100 is 100.0889. Infrared spectrum shows carbonyl absorption at 1712 cm^{-1}.

An aldehyde normally shows a loss of hydrogen (C2 cleavage) from the molecular ion but there is no ion at *m/e* 99 so that compound C is probably a ketone, RCOR′.

An abundant ion at *m/e* 85 must be due to loss of 15 mass units (CH_3) from the molecular ion and suggests C2 cleavage adjacent to carbonyl (R = CH_3):

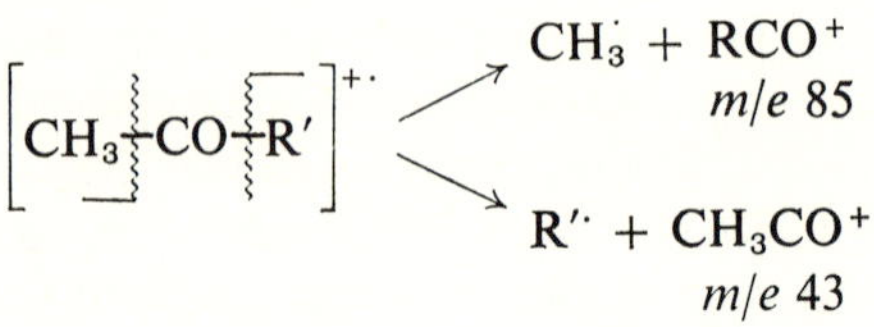

The alternative cleavage adjacent to the carbonyl group would give an ion at *m/e* 43 as shown and the base peak in the spectrum of compound C occurs at *m/e* 43. Since the total composition of C is $C_6H_{12}O$, the group R′ must be C_3H_7 and the possible structures of the unknown have been reduced to two i.e. R′ = n- or isopropyl. To distinguish between the two possibilities, the mass spectrum is examined further and the odd-electron ion at *m/e* 58 noted. This ion arises by the six-centre rearrangement (RE2) of ketones and means that R′ has a γ-carbon atom and so is n-propyl. Compound C is 2-hexanone.

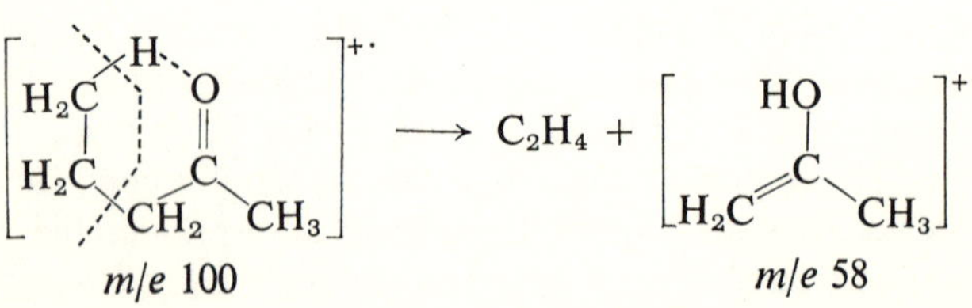

D. The mass spectrum (figure 4.27) is typically 'aromatic' in appearance with an abundant molecular ion and relatively few fragment ions.

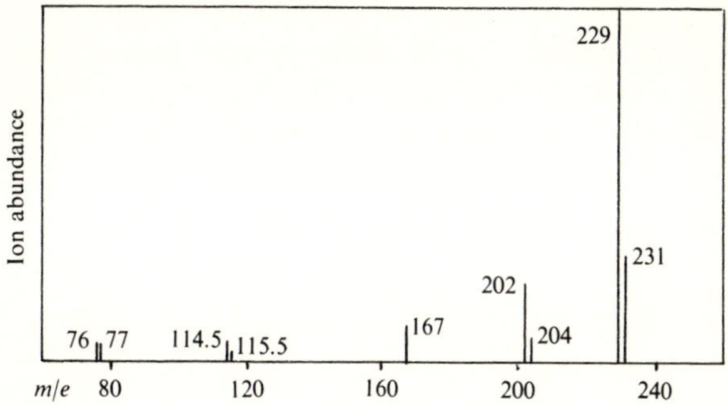

Figure 4.27. Mass spectrum of unknown D. Measured accurate masses of ions are 229.0405, 202.0296. Metastable peaks occur at *m/e* 138.5, 136.0.

In the molecular ion region, there are two peaks at *m/e* 229, 231 separated by two mass units and having an abundance ratio of 3:1 indicating the presence of chlorine. The odd molecular weight suggests the presence of an odd number of nitrogen atoms and this is confirmed by the accurate mass (229.0405) of the molecular ion since the composition $C_{12}H_8ClN_3$ requires a mass of 229.0408. The error between the measured and required masses is 0.003 in 229.0408 i.e. the error is 1.5 p.p.m. and is quite acceptable. Two fragment ions at *m/e* 202, 204 in the ratio 3:1 correspond to the loss of 27 mass units from the molecular ion and suggest the expulsion of C_2H_3 or HCN.

The loss of HCN from aromatic amino-compounds and nitrogen heterocyclic compounds is commonly observed and is confirmed by the accurate mass 202.0296 corresponding to the composition $C_{11}H_7ClN_2$ ($C_{12}H_8ClN_3 - HCN = C_{11}H_7ClN_2$). Metastable ions at *m/e* 138, 136.5 correspond to the losses of ^{35}Cl, ^{37}Cl from *m/e* 202, 204 respectively to yield the ion at *m/e* 167 containing no chlorine (there is no ion at *m/e* 169 of one-third the abundance of the *m/e* 167 ion). The ions at *m/e* 77, 76 are usually observed with aromatic compounds containing benzene rings and the relatively abundant doubly charged ions at *m/e* 114.5, 115.5 suggest more than one aromatic ring is present. Note the molecular formula $C_{12}H_8ClN_3$ which corresponds to ten double bond equivalents and signifies a highly unsaturated or condensed ring system.

Thus, the mass spectrum can be used to propose that the compound is aromatic, possibly a condensed polycyclic with one or more benzenoid rings and contains one chlorine and three nitrogen atoms. The ejection of one nitrogen atom as HCN from the molecular ion before chlorine is ejected strongly suggests that an amino-group is present. Beyond this, further deductions would be speculative unless supported by other experimental information or experience based on similar compounds. The compound is 1-amino-3-chlorophenazine (D).

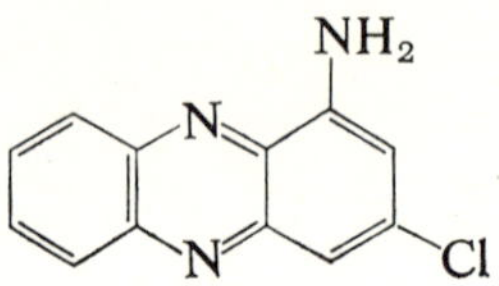

(D)

E. The mass spectrum (figure 4.28) is typically aliphatic with many abundant fragment ions. The molecular ion at *m/e* 228 is relatively abundant for an aliphatic compound and suggests the presence of unsaturation or of one or more ring systems. The ion at *m/e* 210 corresponds to loss of 18 mass units from the molecular ion (metastable ion at *m/e* 193.4) and must be loss of H_2O; a second expulsion of H_2O is apparent from the presence of an ion at *m/e* 192 and a metastable ion at *m/e* 175.5 (= $192^2/210$). The two successive losses of H_2O indicate two hydroxyl groups in compound E. The ion at *m/e* 197 corresponds to ejection of 31 mass units from the molecular ion. This mass loss

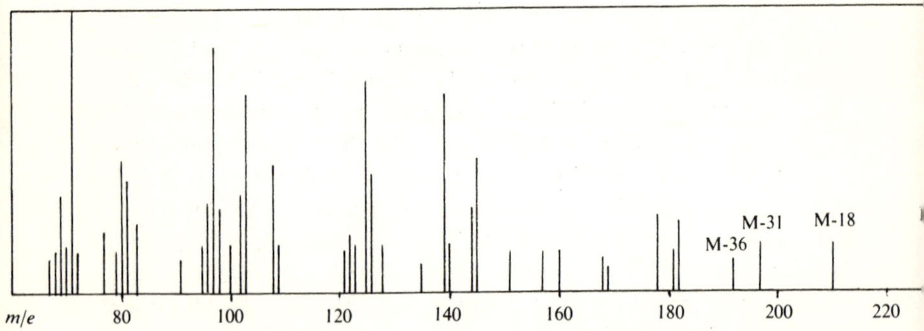

Figure 4.28. Mass spectrum of unknown E. Metastable ions occur at *m/e* 193.4, 175.5. Compositions of ions at *m/e* 228, 210, 197, 192, 169 are respectively $C_{11}H_{16}O_5$, $C_{11}H_{14}O_4$, $C_{10}H_{13}O_4$, $C_{10}H_{12}O_3$, $C_9H_{13}O_3$. Elemental analysis: C, 57.9, H, 7.1 per cent.

frequently corresponds to the scission of CH_3O from a methyl ester; the accurate masses of the ions at *m/e* 197, 228 give a mass difference of 31.0180 mass units (CH_3O requires 31.0184). Confirmation of the ester character is found in the ion at *m/e* 169 of composition $C_9H_{13}O_3$ (loss of CH_3OCO from molecular ion). The accurate mass of the

$$R\cdot CO\cdot OCH_3^{+\cdot} \longrightarrow R\cdot CO^+ \longrightarrow R^+$$

molecular ion yields the molecular formula, $C_{11}H_{16}O_5$. The fragmentation of the molecular ion has indicated that four of the oxygen atoms reside in two hydroxyl and one methyl ester functions. The molecular formula agrees with the elemental analysis (found: C, 57.9; H, 7.1. $C_{11}H_{16}O_5$ requires C, 57.9; H, 7.1 per cent) indicating the absence of non-isomeric impurities. Without experience, it would be difficult to proceed further with just the information from the mass spectrum. The compound is actually the aglucone of loganin (E). In a series of closely related compounds having a similar carbon skeleton, the mass spectra show a prominent ion at *m/e* 139 of composition, $C_7H_7O_3$. The ion is thought to arise from one particular part of the carbon skeleton, as shown, and may be termed a characteristic ion of the series. It is useful because its abundance varies greatly with the presence and position of a double bond in the five-membered ring and so gives added structural information.

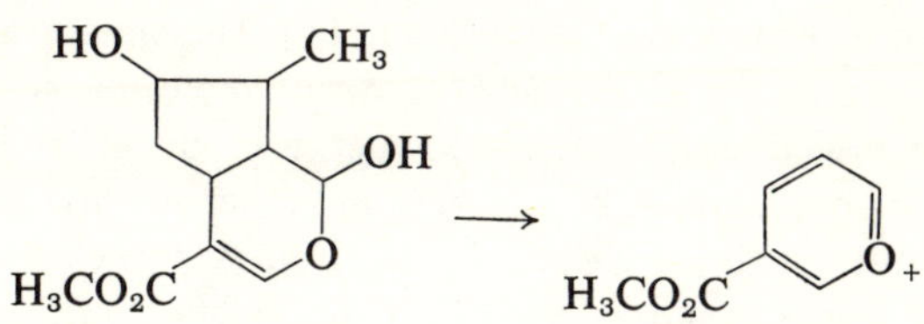

(E) *m/e* 139

F. The mass spectrum (figure 4.29*a*) shows few fragment ions suggesting that compound F may have an aromatic structure. The very abundant ion at *m/e* 91 (C_7H_7) is generally characteristic of benzylic compounds. The low abundances of ions at greater *m/e* values is similar to the behaviour of aliphatic compounds and this evidence taken with the evidence for an aromatic structure indicates that F may be aromatic with an aliphatic or other readily cleaved side-chain. The highest mass ions are those at *m/e* 132, 133 and their abundances are approximately in the ratio of 4:1. At *m/e* 132, the compound could contain a maximum

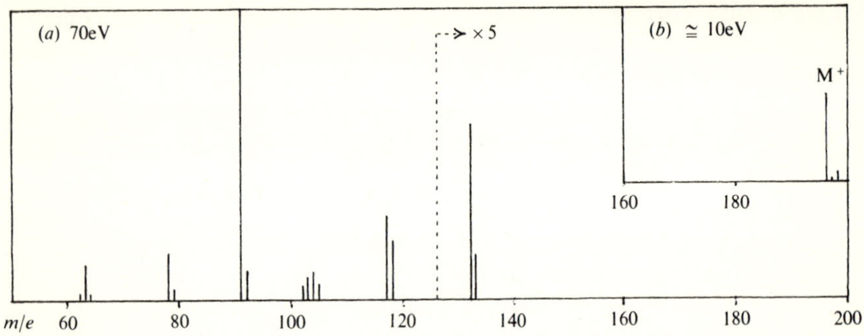

Figure 4.29. Mass spectrum of compound F at (*a*) 70 eV and (*b*) approximately 10 eV (inset for molecular ion region). Composition of molecular ion is $C_{11}H_{16}OS$ and of ions at *m/e* 63, 78 are CH_3OS C_2H_6OS respectively.

of ten carbon atoms and if *m/e* 133 were the M + 1 ion of *m/e* 132, it would have a maximum abundance of approximately $1.1 \times 10 = 11$ per cent of the ion at *m/e* 132. The relative abundance of the ion at *m/e* 133 to that at *m/e* 132 is clearly greater than that, and it cannot be simply the M + 1 ion with *m/e* 132 the molecular ion. Further confirmation of the suspicion that *m/e* 132 is not the molecular ion is found in the presence of ions at *m/e* 117, 118 which would represent losses of 15 and 14 mass units from *m/e* 132. Whereas the loss of 15 mass units (CH_3) is common, the loss of 14 mass units (CH_2) is most unusual; again this suggests that *m/e* 132 is not the molecular ion. The mass spectrum could be very misleading therefore since it shows no molecular ion. To obtain a molecular ion either a low electron energy spectrum, preferably with a cooled ion source, must be run or some other method of ionization used e.g. photo-, field-, or chemical-ionization. At low electron energies, a molecular ion is found at *m/e* 196 (figure 4.29*b*); note the relatively prominent M + 2 ion indicating the presence of sulphur (accurate mass measurement yielded the composition $C_{11}H_{16}OS$). The difference in mass between the true molecular ion at *m/e* 196 and the highest *m/e* value (133) found at 70 electronvolts is 63 mass units and an ion containing sulphur at *m/e* 63 is observed. In other cases where the molecular ion is absent, it may be possible to deduce its *m/e* value by addition of the masses of fragment ions.

Examination of the spectrum reveals an ion at *m/e* 78 (even mass, odd-electron ion) of composition C_2H_6OS and suggests rearrangement has occurred. The compound is methyl 4-phenylbutyl sulphoxide, and the rearrangement process is an example of the very common six-centre process (RE2) discussed in §4.6. Once the compound is known to be a sulphoxide the mass spectrum can aid structure elucidation much more.

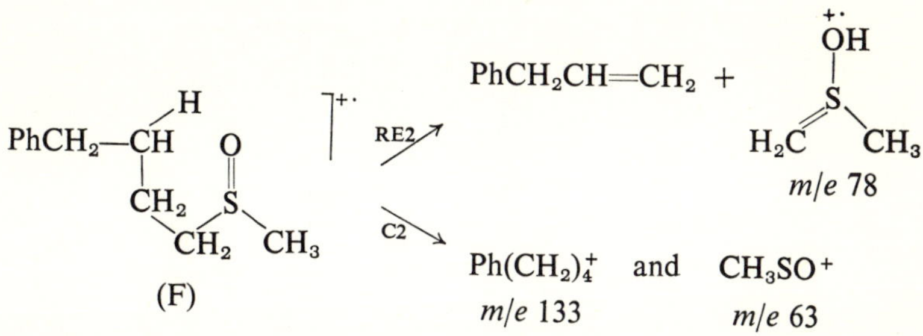

For example, a fragment CH_3SO (63 mass units) is lost from the molecular ion and this unit must be at the end of a chain (C2 cleavage) itself indicating a sulphoxide. The rearrangement reaction implies there must be at least three carbon atoms (six-centre transition state) on the other side of the sulphoxide group. The abundant ion at *m/e* 91 indicates a benzylic grouping. The assembled information leads to the correct structure of compound F after accepting the presence of a sulphoxide group in it.

G. The mass spectrum (figure 4.30*a*) shows groups of peaks at high *m/e* values but only isolated ions at low *m/e* values. There are ten major peaks in the molecular ion region. The last peak at *m/e* 305 must be a ^{13}C isotope leaving nine other possible isotope peaks in the ratio 1.0:0.2:0.6:1.1:1.2:0.7:1.6:0.3:0.7. An examination of isotope tables suggests molybdenum is present since it has seven isotopes at 92, 94, 95, 96, 97, 98, 100 in the ratio 1:0.6:1.0:1.0:0.6:1.5:0.6 (the peaks at *m/e* 297, 303 would then have to be ^{13}C isotope peaks of *m/e* 296, 302 respectively). Confirmation of this identification is obtained from the presence of ions at *m/e* 92, 94, 95, 96, 97, 98, 100 albeit in low abundance. If the peak at *m/e* 297 is a ^{13}C isotope contribution, then the abundance of this ion relative to that of *m/e* 296 enables the ^{13}C isotope contributions to be removed from all the ions between *m/e* 296 and *m/e* 305. When this is done, the residual peaks number only seven and are in the ratio 1.0:0.6:1.0:1.1:0.7:1.5:0.7 i.e. very close to the ratios required for molybdenum. All isotope contributions other than ^{92}Mo and ^{12}C can then be removed from the spectrum leaving it much simpler to interpret (figure 4.30*b*).

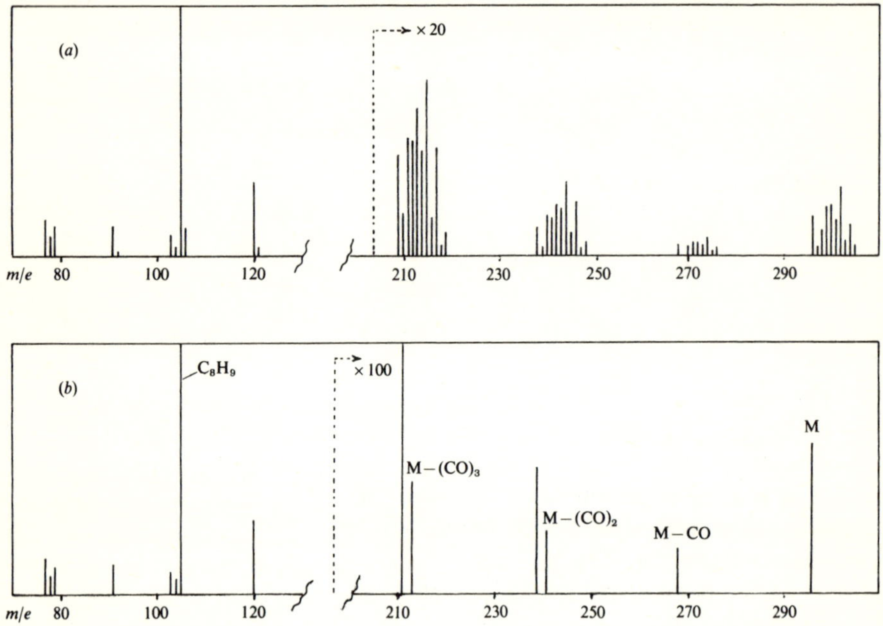

Figure 4.30. (*a*) Mass spectrum of unknown G; (*b*) the same mass spectrum with isotope contributions removed. Metastable peaks appear at *m/e* 236, 208, 101, 91.8. Very low abundance ions also occur at *m/e* 92, 94, 95, 96, 97, 98, 100. The ion abundances above *m/e* 200 are drawn to increased scales as shown.

Successive losses of 28 mass units are observed in the presence of peaks at *m/e* 296, 268, 240, 212 (on the evidence of metastable ions, the ions at *m/e* 238, 210 arise from *m/e* 240, 212 respectively by loss of H_2). Three successive ejections of 28 mass units from an organometallic compound suggest elimination of three CO molecules from a metal complex (confirmed by accurate mass measurement). Thus, the compound contains molybdenum (^{92}Mo, figure 4.30*b*) and three CO ligands ($3 \times CO = 84$ mass units) and together these account for $92 + 84 = 176$ mass units leaving $296 - 176 = 120$ mass units. An abundant ion is found at *m/e* 120 and decomposes by elimination of 15 mass units (CH_3) to give the ion at *m/e* 105 (metastable peak for process at *m/e* 91.8).

Accurate mass measurement shows that the ion at *m/e* 105 which also loses H_2, has the composition C_8H_9. Apart from molybdenum and the CO groups, the remainder (C_9H_{12}) of compound G must be very unsaturated (four double bond equivalents) and together with ions at *m/e* 77, 78, 79, 91 this suggests a benzenoid ring, possibly benzylic, with a C_3H_7 side-chain. The prominent loss of methyl from *m/e* 120 and the relatively small peak at *m/e* 91 indicate that the side-chain is isopropyl. In this instance, on the basis of the mass spectrum alone, a tentative total structure can be proposed for compound G viz. $Mo(CO)_3$ complexed to isopropylbenzene. The actual structure of (G) is as shown.

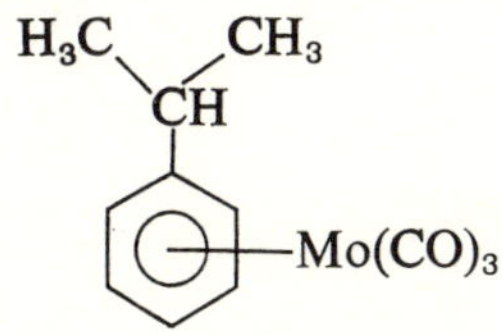

(G)

H. Analysis of a peptide showed that the constituent amino-acids were glycine, alanine, and phenylalanine. The peptide was acetylated at the *N*-terminal and methylated at the carboxyl end. The volatility of a peptide can be increased by *N*-methylation which removes amide

$$H_2N\sim\sim CO_2H \longrightarrow CH_3CONH\sim\sim CO_2H \longrightarrow CH_3CONH\sim\sim CO_2CH_3$$

hydrogen bonding. The acetylated methyl ester was accordingly methylated at each amide nitrogen using sodium hydride/dimethylformamide/methyl iodide. A partial mass spectrum of the completely methylated peptide is presented in figure 4.31. Table 4.10 represents

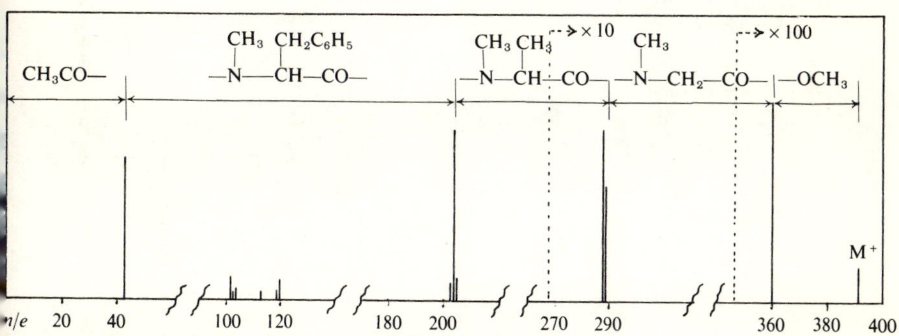

Figure 4.31. Partial mass spectrum of peptide, H. See table 4.10 for accurate masses of ions.

TABLE 4.10 *Partial output of accurate masses of ions in mass spectrum of peptide* (H) *from mass spectrometer–computer data acquisition system*

Accurate mass	Ion abundance	Accurate mass	Ion abundance
391.20840	5	130.05032	20
361.18626	4	130.04034	2
360.18897	32	129.09114	12
346.17189	7	129.05521	3
300.15618	101	128.07452	3
290.15678	27	128.06871	7
289.15330	167	127.14871	6
288.14672	237	127.08769	97
287.14343	3	127.07702	5
286.14076	12	127.06245	6
276.14373	7	127.05297	2
275.13824	47	126.05695	10
274.13222	59	125.13269	14
262.16439	14	125.05918	2
261.16113	62	123.11701	10
261.12704	2	121.10190	5
260.11448	5	121.08417	23
218.11563	19	121.06390	10
217.13195	7	120.08137	278
217.10954	64	119.08703	5
216.10224	77	119.07337	188
215.10053	138	118.08704	12
213.08962	6	118.07648	3
210.07113	1	118.06492	40
206.10762	28	117.07983	2
205.10511	331	117.07371	6
205.08560	2	117.06918	7
204.10226	2414	117.05726	32
204.08297	9	116.07129	47
203.13698	3	116.06356	2
203.11933	3	116.04886	11
203.09404	208	115.08686	6
202.08568	22	115.07847	3
201.08531	8	115.05475	14
200.13832	3	114.05527	18
198.09589	2	113.13443	5
197.09247	14	113.13013	3
192.10293	6	113.07191	45
130.08844	11	113.05985	4
130.06529	17	113.04804	56

part of a computer output listing the accurate masses of all ions found by using a mass spectrometer–computer data processing system. From this information, the structure of the peptide can be determined as follows (note, it is not necessary to have an amino-acid analysis for the ensuing structure elucidation, but having one reduces the amount of work and can eliminate ambiguities in interpretation). After methylation, the masses of the amino-acid and acetyl and methoxyl groups are as shown (4.40) Beginning at one end of the peptide with acetyl, add

Residue		*Composition*	*Mass*
N-methylglycyl	$—N(CH_3)—CH_2—CO—$	C_3H_5NO	71.0371
N-methylalanyl	$—N(CH_3)—CH(CH_3)—CO—$	C_4H_7NO	85.0528
N-methylphenyl-alanyl	$—N(CH_3)—CH(CH_2C_6H_5)—CO—$	$C_{10}H_{11}NO$	161.0841
Acetyl	$CH_3CO—$	C_2H_3O	43.0184
Methoxyl	$CH_3O—$	CH_3O	31.0184

(4.40)

to its mass (43.0184) the mass of each amino-acid residue in turn; this gives the results presented in (4.41). An abundant ion of mass 204.1023

Acetyl + *N*-methylglycyl = 114.0555
Acetyl + *N*-methylalanyl = 128.0711 (4.41)
Acetyl + *N*-methylphenylalanyl = 204.1024

is found in the spectrum (very low abundance ions occur at *m/e* 114, 128 but their accurate masses are not those required) and the first amino-acid in the sequence is therefore phenylalanine. Scheme (4.42) shows the masses expected from adding a second residue to phenylalanine. A prominent peak is found at 289.1533 and alanine must

Acetyl + *N*-methylphenylalanyl + *N*-methylglycyl = 275.1396
Acetyl + *N*-methylphenylalanyl + *N*-methylalanyl = 289.1552
(4.42)

follow phenylalanine in the sequence. The final amino-acid residue must be therefore the remaining glycyl, and to complete the sequence the terminal methoxyl group must be added (4.43).

Acetyl + *N*-methylphenylalanyl + *N*-methylalanyl +
N-methylglycyl ⁞ + methoxyl ⁞ (4.43)
(360.1923) ←-- (391.2107) ←--

Ions of the expected accurate masses are found at *m/e* 360, 391 to complete the sequence and the original must have been a tripeptide, Phe·Ala·Gly. A complete structure of the peptide would require knowledge of the D or L nature of the constituent amino-acid residues, but the mass spectrum has yielded the important sequence information. Sequence analysis in this way has now been automated by computer interpretation of the accurate mass data.

$H_2NCHCONHCHCONHCH_2CO_2H$
with side chains $CH_2C_6H_5$ (first CH) and CH_3 (second CH)

(H)

5 Appendix

5.1. Metastable ions. The term 'metastable' has been applied to those ions in a mass spectrometer which, at the detector, have less than the full kinetic energy originally imparted at the ion source. To understand the origin of metastable ions, we may consider a double focussing arrangement as depicted in figure 5.1, and a mass spectrometric reaction

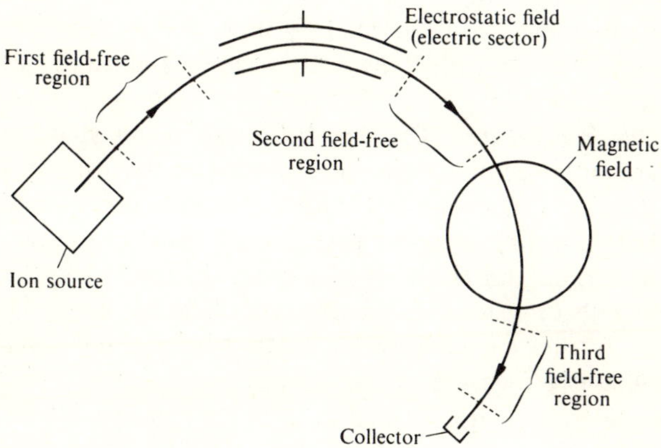

Figure 5.1. Path of main ion beam through mass spectrometer showing the first, second, and third field-free regions.

in which an ion M^+ of mass m_1 yields a fragment ion A^+ of mass m_2 and a neutral part m^0. As was shown earlier, ions M^+, A^+ will be

$$\underset{(m_1/e)}{M^+} \longrightarrow \underset{(m_2/e)}{A^+} + \underset{(m_1 - m_2)}{m^0} \qquad (5.1)$$

present in the ion source and, after acceleration from there, each will possess translational energy equal to $e.V$ where V is the accelerating voltage. By suitably adjusting the magnet current, ions M^+, A^+ are collected at the detector according to the formula, $m/e = H^2r^2/2V$;

these may be called 'normal' ions. Suppose however that some ions (M^+) fragment to give ions (A^+) *after* leaving the ion source and *before* detection. Then, the translational energy in M^+, namely $e.V$, must be shared between A^+ and m^0 in accordance with the law of conservation of momentum. Thus, these 'metastable' A^+ ions will have only a part of the original energy $e.V$, and will not be collected at the detector at the same position as 'normal' A^+ ions having the full translational energy, $e.V$. The metastable A^+ ions have therefore the same mass as normal A^+ ions but are collected differently because they do not have the same translational energy i.e. if the new translational energy is equivalent to V' (where $V' < V$), then $m/e = H^2r^2/2V'$. Since m/e is the same for normal and metastable ions and $V' < V$, a weaker magnetic field is required to focus the metastable ions. The actual place in the mass spectrometer where the loss of translational energy occurs i.e. where the reaction $M^+ \rightarrow A^+ + m^0$ occurs, determines how the metastable ions may be detected. It is necessary to consider distinct areas which may be called the first, second, and third 'field-free' regions (figure 5.1).

First field-free region. This lies between the ion source and the electrostatic analyser of the double focussing instrument; it does not exist in the single focussing instrument. Any A^+ ions formed in this region have the wrong energy to pass through the electrostatic analyser (electric sector) because the latter is an energy focussing device designed to transmit only those ions of the correct translational energy. Therefore, metastable A^+ ions having only a fraction of the full translational energy will not be focussed by the analyser and will not be observed in the mass spectrum. If it is desired to investigate metastable transitions in the first field-free region it is necessary to alter the accelerating voltage at the ion source or the voltages on the electrostatic analyser so as to allow the passage of the metastable ions. If an ion M^+ of mass m_1 after acceleration through a potential (V) decomposes to give a metastable ion A^+ of mass m_2 and a neutral particle m^0 of mass $= (m_1 - m_2)$, the initial translational energy ($e.V$) of m_1 must be shared between m_2 and $m_1 - m_2$. If the accelerating voltage is increased to V' without

$$e.V = \frac{m_2}{m_1}e.V + \frac{m_1 - m_2}{m_1}e.V$$

changing the electrostatic analyser voltage such that $V' = (m_1/m_2)V$, then the translational energy of the metastable ion (m_2) becomes $(m_1/m_2)e.(m_2/m_1)V = e.V$ i.e. it then has the correct translational energy to pass through the electrostatic analyser and be collected. Changing the voltage at the analyser in this way then stops the normal ions getting through; the normal ions are defocussed and hence the term *defocussing technique* for this method of investigating metastable transitions (Barber and Elliott, 1964; Futrell, Ryan, and Siek, 1965; Jennings, 1965). Perhaps it ought to be termed a focussing technique since the metastable ions being investigated are focussed and the normal ions defocussed. This method for examining metastable transitions early along the flight-path of the ions is extremely useful for those cases where metastable ions are not observed in the routine mass spectrum (see second field-free region), or where the metastable peaks lie partly hidden under intense normal ion peaks.

In using this method, suppose a normal ion A^+ is selected and that A^+ may be formed from any or all of the precursor ions M_1^+, M_2^+, M_3^+, M_4^+ (figure 5.2*a*). It is required therefore to look for possible meta-

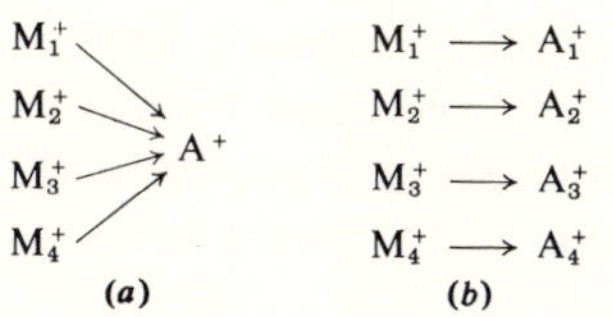

Figure 5.2. Fragment ion A^+ is formed by decomposition of any or all of ions M_1^+, M_2^+, M_3^+, M_4^+. To find which fragmentation path(s) is followed, a search is made for metastable ions A_1^+, A_2^+, A_3^+, A_4^+. (*a*) Possible precursors M_1^+, M_2^+, M_3^+, M_4^+, of normal fragment ion A^+. (*b*) Possible fragmentations in first field-free region giving metastable ions A_1^+, A_2^+, A_3^+, A_4^+.

stable ions A_1^+, A_2^+, A_3^+, A_4^+ (figure 5.2*b*). After selecting the ion A^+, the accelerating voltage is gradually increased whilst maintaining a constant voltage on the electrostatic analyser to focus the possible metastable ions A_1^+, A_2^+, A_3^+, A_4^+. If any of these are detected at the collector then the corresponding fragmentation must occur. The process has been semi-automated and the metastable ion peaks can be recorded directly onto chart paper. As an example, some of the metastable transitions in the mass spectrum of *N*-phenylsuccinimide found by defocussing are shown in figure 5.3; metastable ions observed in the routine spectrum are indicated. It can be seen that more transitions are found by the defocussing method than appear in the routine spectrum. The defocussing technique gives more definitive information than the metastable peaks appearing in a routine spectrum (§1.4). The positions of the latter peaks depend on the parent (m_1) and daughter (m_2) ions

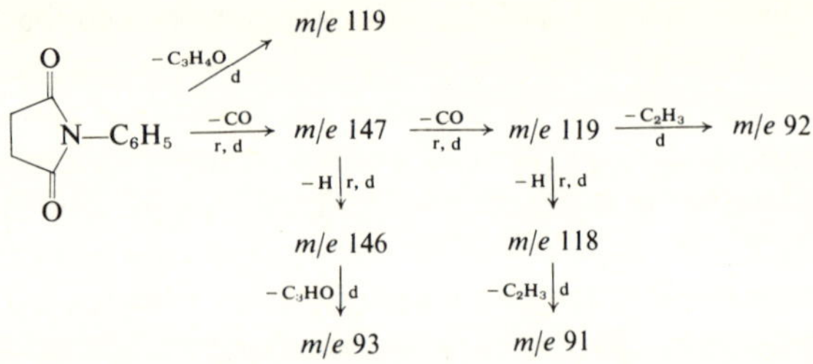

Figure 5.3. Some metastable transitions observed by the defocussing technique (d) and in the routine (r) mass spectrum of *N*-phenylsuccinimide.

which may not be exactly definable in a routine spectrum containing many ions i.e. there can be ambiguity as to which ions are the parent and daughter. In the defocussing method, daughter ions are examined individually and the parent ions giving rise to them are found precisely.

An alternative method of focussing the metastable ions is to keep the ion accelerating voltage constant and vary the electric sector (electrostatic analyser) voltage. During routine operation of the mass spectrometer, ions are focussed at the slit placed between the electric sector and the magnetic field (figure 5.4) but metastable ions are not

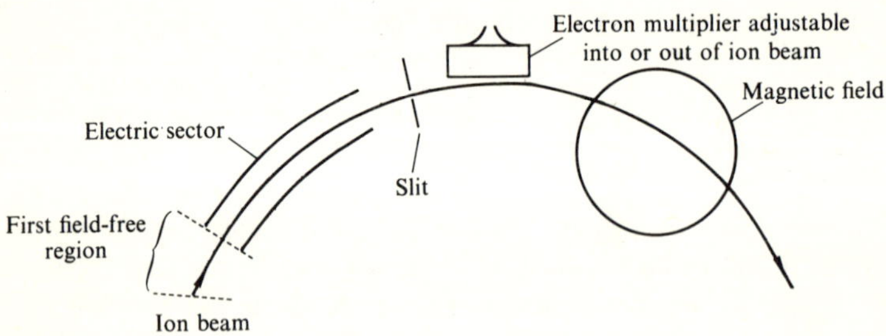

Figure 5.4. Location of electron multiplier behind slit after electric sector for use in ion kinetic energy spectroscopy.

focussed. By altering the electric sector voltage, the normal ions are defocussed but the metastable ions can then be focussed. This method of examining metastable ion formation has been termed ion kinetic energy spectroscopy (IKES; Beynon *et al.* 1970). A second electron multiplier which can be raised or lowered into place behind the slit (figure 5.4), and an arrangement for continuously varying the voltage on the electrostatic analyser plates without affecting the ion accelerating voltage are the extra requirements for using this method.

In general an ion m_1 with x positive charges may decompose to an ion m_2 with y positive charges in the first field-free region either by a 'unimolecular' reaction or as a result of collision with a neutral particle. Any ion passing through the electric sector is constrained into a curved path of radius (r) given by, $r = 2V/B$ (see equation 2.1) where B is the electric sector voltage, and V the ion accelerating voltage. An ion m_1 with x positive charges on decomposition in the field-free region produces the ion m_2 (y positive charges) with kinetic energy equivalent to $(m_2/m_1).x.V$ rather than V (see above). The radius of curvature (r_1) of the ion m_2 in the electric sector is given by equation (5.2). Since m_2 has y positive charges, the effect of the potential B is magnified

$$r_1 = \left(\frac{m_2}{m_1}x\right)\frac{2V}{B} \tag{5.2}$$

y times above that of a singly charged ion so that the radius of curvature (r_2) must be corrected for this as shown in equation (5.3). Therefore to constrain the ion m_2 into a radius of curvature (r) for focussing,

$$r_2 = \left(\frac{m_2}{m_1}.\frac{x}{y}\right)\frac{2V}{B} \tag{5.3}$$

the radius (r_2) must be multiplied by the factor $(m_1/m_2).(y/x)$. Changing the electric sector voltage from B to $(m_2/m_1).(x/y)B$ will change the radius of curvature of the ion m_2 from r_2 to r and focus it at the slit. Examination of the expression for the required voltage, $(m_2/m_1).(x/y)B$ leads to a number of possible results. If $x = 2y$, and $m_2 = m_1$ i.e. for a collision reaction in which the process $M^{++} \rightarrow M^{+}$ occurs, then $(m_2/m_1).(x/y)B = 2B$; the ions m_2 appear at twice the normal electric sector voltage, B. If the doubly charged ion, M^{++} decomposes to two singly charged ions, P^{+} and Q^{+}, then $x = 2y$ and $(m_2/m_1).(x/y)B$ becomes $(P/M)2B$ or $(Q/M)2B$ where P, Q, M are the masses of ions P^{+}, Q^{+}, M^{++}, i.e. there is a metastable ion peak at each of these voltages. An example of this behaviour can be seen amongst the metastable ions in the spectrum of naphthalene. The doubly charged ion $C_{10}H_6$ (126 mass units) fragments to yield two singly charged ions, C_7H_3 (87 mass units) and C_3H_3 (39 mass units). The metastable ions occur at electric sector voltages of $1.38B$ (=[87/126].$2B$ and $0.62B$

$$\underset{(126)}{C_{10}H_6^{++}} \longrightarrow \underset{(87)}{C_7H_3^{+}} + \underset{(39)}{C_3H_3^{+}}$$

(= [39/126].$2B$) respectively. When $x = y$, the required sector voltage for focussing ($[m_2/m_1]B$) is less than B since m_2 is less than m_1. Thus metastable ions may be observed at electrostatic analyser voltages below or above the normal operating voltage B without interference from normal ions. The extra ion collector placed behind the slit (figure 5.4) enables focussing of the metastable ions to be carried out by adjusting the electric sector voltage continuously. If the ion collector is raised out of the ion beam, the ions pass on and into the magnetic analyser for mass analysis. The mass analysis is necessary to discover which metastable ions are being investigated. In the example given above of the ion $C_{10}H_6^{+\,+}$ decomposing to give ions at $C_7H_3^+$ and $C_3H_3^+$, mass analysis of the metastable ions found at a sector voltage of $1.38B$ showed that they had a mass-to-charge ratio of 87 i.e. $m_2 = 87$. Thus, $1.38B = (87/m_1)2B$, from which $m_1 = 126$; mass analysis of the metastable ions focussed at $1.38B$ gave the daughter ion and this combined with the electric sector voltage gave the parent ion. The ion at mass 126 must have been the progenitor of the ion at mass 87.

By increasing the 'back-ground' gas pressure in the first field-free region, collision induced decompositions of ions can be investigated by observing the changes produced in metastable ion abundances.

Second field-free region. This region extends from the end of the electrostatic analyser up to the beginning of the magnetic field. In a single focussing instrument, it is the space between the ion source and the magnetic field. Because decompositions in the second field-free region occur after the electrostatic analyser, there is no discrimination against the metastable ions which are mass analysed in the magnetic field and appear in routine spectra from both single and double focussing instruments as broader peaks than those from normal ions (§1.4). The focussing in the magnetic field is affected by both the m/e value of an ion and its translational energy and, therefore, the metastable A^+ ions will be collected differently from normal A^+ ions. The metastable A^+ ions have less translational energy and are deflected more in the magnetic field than normal A^+ ions.

There is a simple formula (5.4) connecting the apparent mass (m^*) of the metastable ion with the masses of the precursor ion (m_1) and fragment or daughter ion (m_2). If a metastable ion is observed at m^* in the routine mass spectrum, it is only necessary to apply the formula

$$m^* = m_2^2/m_1 \qquad (5.4)$$

(5.4) to determine the precursor and fragment ions it arises from. For example, the mass spectrum of toluene contains large peaks at *m/e* 91, 65 and a strong metastable ion peak at *m/e* 46.4. Since $65^2/91 = 46.4$, we can say that some of the ions at *m/e* 65 arise from ions at *m/e* 91 by ejection of 26 mass units.

If there are many peaks in the mass spectrum, it is time consuming to check all the possible precursor and fragment ions which could give rise to an observed metastable ion, and so tables and nomograms (McLafferty, 1966) are available to speed the checking. Anyone with a minimum of computer programming ability and access to a computer will find it a trivial exercise to program a computer to do this checking.

Ususally, metastable ion peaks have roughly Gaussian shapes in the mass spectrum with frequently a 'tail' on the high mass side (figure 5.5*a*). Sometimes the metastable ions appear as broad, flat peaks extending over several mass units (figure 5.5*b*,*c*,*d*); this arises

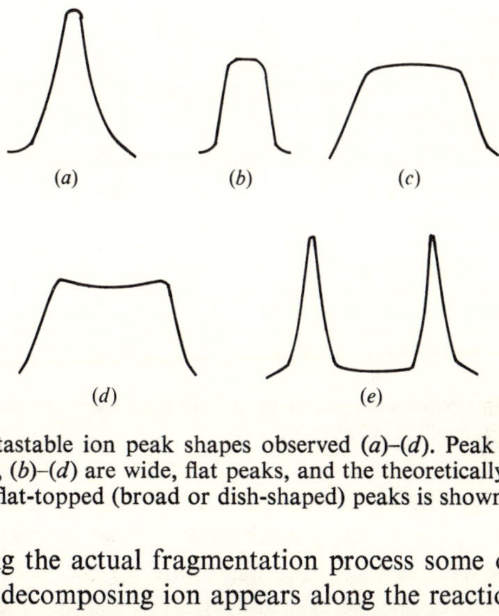

Figure 5.5. Metastable ion peak shapes observed (*a*)–(*d*). Peak (*a*) is the shape normally found, (*b*)–(*d*) are wide, flat peaks, and the theoretically expected shape at one of these flat-topped (broad or dish-shaped) peaks is shown in (*e*).

because during the actual fragmentation process some of the internal energy in the decomposing ion appears along the reaction co-ordinate as kinetic energy. Figure 5.5*e* shows the theoretically expected peak shape which is rarely observed normally because energy spread in the ion beam causes filling of the region between the two sharp peaks. Theoretically expected peak shapes can be observed by using slits to cut down the spread of the ion beam. The energy (T) released in giving rise to the broad metastable ions may be calculated from the formula

$$T = \left(\frac{d.m_1}{4\,m_2^2}\right)^2 \left(\frac{e.V}{\mu}\right) \qquad (5.5)$$

(5.5) in which d is the width of the peak in mass units, m_1 and m_2 are the masses of the precursor and fragment ions, and μ is a reduced mass $(m_1 - m_2)/m_2$ (Beynon *et al.*, 1968).

There is another method of examining metastable ions from the second field-free region. Since metastable ions have less translational energy than normal ions, the two types may be differentiated by having an ion repeller voltage at the collector. Suppose normal ions possess translational energy corresponding to V volts, and metastable ions V' volts. If the repeller voltage (V_r) is greater than V (and therefore V') all ions are turned back and strike a metal plate causing secondary electrons to be emitted (figure 5.6). These secondary electrons are

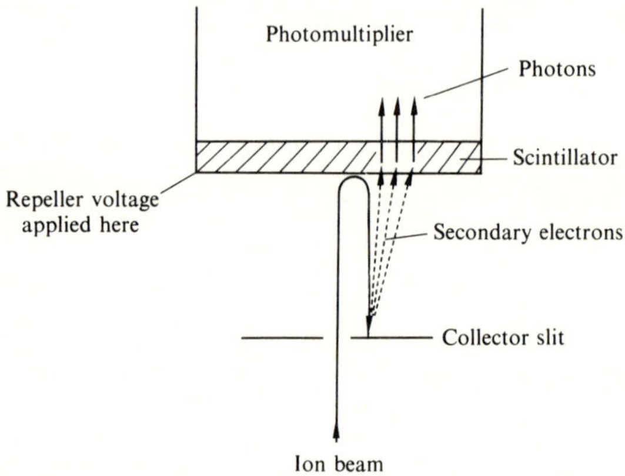

Figure 5.6. Arrangement of scintillator–photomultiplier and ion beam repeller voltage for detection of metastable ions.

detected by a scintillator–photomultiplier. If now the repeller voltage (V_r) is made less than V but greater than V' then normal ions are not repelled but metastable ones are, again causing emission of secondary electrons. Therefore, the method allows a normal mass spectrum to be obtained at one voltage $V_r(> V)$ and then, after changing the repeller voltage to $V > V_r > V'$ a second scan records only the metastable ions arising from transitions in the second and third field-free regions. Like the defocussing technique, this method shows metastable ion peaks which may be obscured by normal ion peaks in a routine mass spectrum.

Third field-free region. After the end of the magnetic field and before the collector there is a small region (figure 5.1) where metastable decompositions may take place. The region lies after all focussing has occurred so there is no separation of normal and metastable ions which arrive at the collector at the same m/e value i.e. $m^* = m_1$. However, metastable A^+ ions still possess less translational energy than normal A^+ ions and a collector with a repeller voltage can discriminate between them. The metastable ion detector with repeller electrode described at the end of the section on the second field-free region is suited to this end and does in fact show metastable transitions occurring in both the second and third regions. Metastable ions originating in the third field-free region appear as sharp peaks in this method since no mass separation occurs after the end of the magnetic field.

Other regions. The other places along the flight path of the ions between the source and the collector where metastable ions may be formed lie in magnetic or electric fields and observing them is not so simple. At the beginning of the magnetic field (end of second field-free region), the relation $m^* = m_2^2/m_1$ holds but at the end of the magnetic field (beginning of the third-free region), $m^* = m_1$. Inside the magnetic field therefore any metastable ions formed will be collected between m_2^2/m_1 and m_1 and this leads to a long tail appearing on the high mass side of the routinely observed metastable ion peak and extending up to the precursor ion peak (figure 5.7). Because metastable

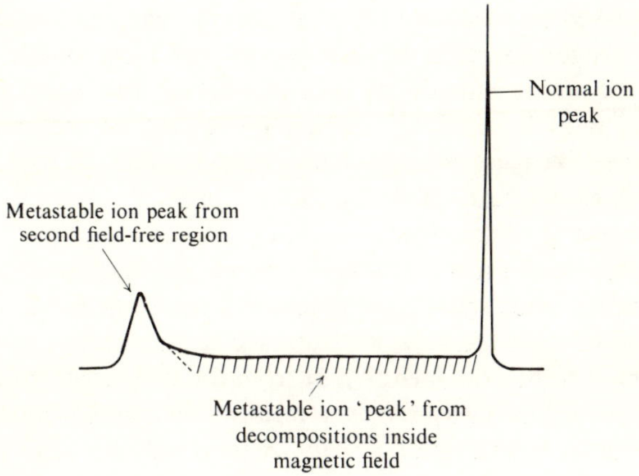

Figure 5.7. Appearance of metastable ions formed in magnetic field.

ion peaks are, in any case, not intense, this diffusion of the peaks into a long tail is infrequently noticeable at the usual amplification required for recording a normal mass spectrum.

Metastable ions formed in the electrostatic analyser region are not generally readily observable.

5.2 Accurate mass measurement. To calculate the elemental composition of an ion its mass must be measured correct to three or four decimal places. Such accurate mass measurement necessitates the use of a double focussing instrument or, in a more limited way, a very good single focussing machine. The measurements are carried out either manually or with a computer but in both methods the accurate mass of an ion is obtained by comparing its mass with that of an ion of accurately known mass. For example, if the reference ion has a mass of 68.99521 and the mass ratio between it and the ion of unknown mass is 1.203520 then the unknown mass is $1.203520 \times 68.99521 = 83.037111$ mass units. By calculation it is found that the composition C_4H_5NO corresponds to this mass (see table 1.4; $C_4 = 48.00000$, $H_5 = 5.03910$, $N = 14.00307$, $0 = 15.99491$, total $= 83.03708$ mass units).

Manual measurement. A reference compound, usually a perfluorinated hydrocarbon, and the compound being investigated are put into the mass spectrometer at the same time. Suppose it is required to measure the mass of an ion M^+, then from the reference compound is selected an ion R^+, of mass close to that of M^+. The peak from ion R^+ is displayed on a cathode ray tube and, by changing the accelerator voltage at the ion source in small steps, the peak from ion M^+ can also be displayed on the cathode ray tube alternately with R^+. The voltage is changed until R^+ and M^+ appear coincident on the cathode ray oscilloscope screen; the voltage change gives directly the required mass ratio which can be read off the instrument panel.

The method just described is called *peak matching* and to obtain accurate mass ratios it is essential to operate the mass spectrometer at high resolution when small mass differences can be detected. The manual method of mass measurement is time consuming when more than a few ions have to be examined. Apart from the time itself, the relatively slow nature of the measurements means that gas chromatographic effluents cannot be examined directly. Also, if only a small quantity of material is available, as with many natural products, slow accurate mass measurement may consume an unacceptably high proportion of it. For these reasons, computers have been introduced to enable the necessary information for mass measurement to be gathered and processed very quickly.

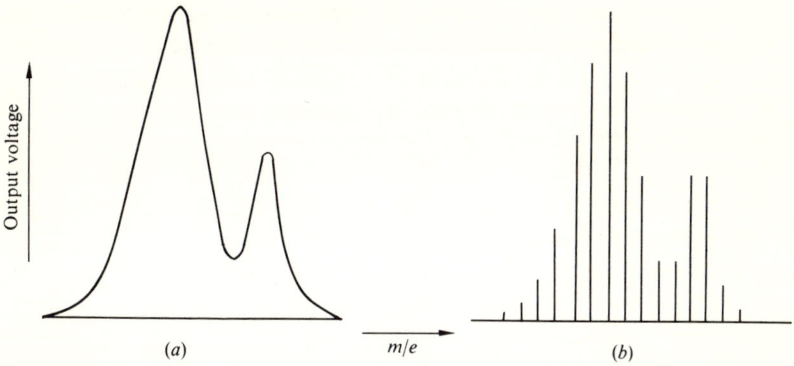

Figure 5.8. Analogue signal (*a*) and its digitized form (*b*) from a mass spectrometer.

Computer measurement. The electrical signal from the mass spectrometer is a continuously varying one (figure 5.8*a*), and so each peak is part of an analogue signal. The analogue signal may be processed directly using an analogue computer, but most computers are digital devices which require the incoming signal to be in digital form (figure 5.8*b*). In these cases an analogue to digital converter must be placed between the mass spectrometer and the computer to digitize the incoming signal. In whichever way the information is fed to the computer, it consists essentially of the peaks due to both reference compound and the compound being examined. The computer is programmed to find the reference peaks and then to calculate the accurate masses of the ions from the compound investigated by interpolation between the reference masses. The computer can then give this information as a simple listing of accurate masses and intensities or it can calculate the elemental compositions directly. Table 5.1 shows some typical results from part of a computer examination of the mass spectrum of methyl stearate. The table records the actual measured mass, the intensity of the peak in arbitrary units, possible elemental compositions for the ion, and the error between the measured mass and that calculated for the elemental composition.

The computer and mass spectrometer may be linked on-line or off-line. In the on-line mode, signals are taken directly from the mass spectrometer into the computer and processed immediately. In the off-line mode, the signals are stored first on magnetic tape and processed

TABLE 5.1 *Computer print-out of measured masses and elemental compositions of some ions in the spectrum of methyl stearate*

Measured mass	Peak intensity	Elemental composition							Error in mass
		$^{12}C/^{13}C$	H	N	O	Others			
300.2975	133	18/1	39	0	2	0	0	0	−0.9
299.2919	835	18/1	38	0	2	0	0	0	1.4
†298.2883	4919	19/0	38	0	2	0	0	0	1.1
271.2610	290	17/0	35	0	2	0	0	0	−2.7
		16/1	34	0	2	0	0	0	1.7
270.2558	1819	17/0	34	0	2	0	0	0	−0.1
269.2474	1222	17/0	33	0	2	0	0	0	−0.6
‡268.2835	59	No composition calculated							
268.2716	377	17/1	35	0	1	0	0	0	−0.6
‖267.2687	1745	18/0	35	0	1	0	0	0	−0.1
256.2358	1070	15/1	31	0	2	0	0	0	0.0
255.2306	4734	16/0	31	0	2	0	0	0	−1.8
		15/1	30	0	2	0	0	0	2.6

† This is the molecular ion, $^{12}C_{19}H_{38}O_2$; the measured mass at 298.2883 differs from the mass calculated for this composition by 1.1 millimass units. Note the M + 1 ion at 299.2919 of composition, $^{13}C_1{}^{12}C_{18}H_{38}O_2$.

‡ This is a small peak in the mass spectrum for which no composition could be calculated from the given parameters. It probably represents a small amount of impurity.

‖ This ion, $^{12}C_{18}H_{35}O$, arises by loss of CH_3O from the molecular ion.

separately on a computer when convenient. Mass spectra may be scanned at high resolution in 10 seconds or less so that the computer-aided method of mass measurement might be used with gas chromatographic effluents and for small quantities of material.

In the photographic plate method of recording a mass spectrum, there are no electrical signals from the mass spectrometer to pass to a computer. The record of the mass spectrum on the photographic plate may be converted into an analogue electrical signal by using a microdensitometer. A narrow light beam is passed through the photographic plate (figure 5.9) and is collected at a detector which puts out an electrical signal proportional to the amount of light falling on it. The photographic plate is passed at a constant speed through the light beam and the images on the plate cause fluctuations in the detected light level; the analogue electrical signal from the detector may be passed directly to a computer as described above. The method inevitably involves off-line working of the computer.

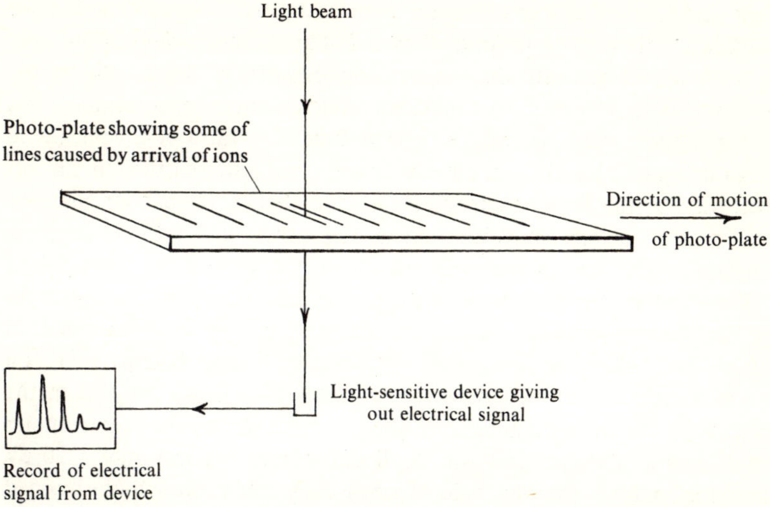

Figure 5.9. Diagrammatic representation of the action of an optical microdensitometer.

5.3. Combined gas chromatographic–mass spectrometric methods. A mass spectrometer normally requires samples to be in the vapour state before ionization; gas chromatographic apparatus provides samples as effluents in a gas stream which are already vaporized. It is not surprising to find these two powerful analytical techniques combined in one unit. In such a g.c.–m.s. combination, two major difficulties must be overcome and arise from the g.c. carrier gas itself and the times taken for substances to emerge from the g.c. column.

Carrier gas. The flow of gas through analytical g.c. columns varies from about 0.2–1.0 ml/min for capillary columns to about 50 ml/min for packed columns. Except for the very slowest gas flows through a capillary column, the pumps maintaining a low pressure in the ion source are unable to cope with these rates of flow and the ion source pressure would quickly rise to unacceptable values outside normal working limits. Therefore, the flow of carrier gas into the ion source must be restricted either by allowing only a small fraction of the total g.c. effluent into the source (splitting) or by first separating the carrier gas from the organic effluent. Whether splitting or separation is used, the aim is to transfer a minimum of carrier gas to the ion source but a maximum of entrained organic matter.

The simplest technique is to use a 'splitter' between the g.c. column exit and the ion source entrance. For example if the gas flow from a capillary column is 1 ml/min, then a 1:4 split ratio would allow only 0.2 ml/min of gas into the source, and this can be dealt with by the pumps. Four-fifths of the material eluted from the column is not utilized and passes out into the atmosphere. The ratio of carrier gas to organic material in the g.c. effluent is high, and much of the ion current recorded by the mass spectrometer is generated by the carrier gas. For this reason, helium may be used as the carrier gas and the ionizing energy in the ion source reduced below the ionization potential of helium but kept above the ionization potentials of organic compounds. In an electron-impact source, the electron energy is maintained at 20 eV i.e. below the ionization potential of helium (24.581 eV). Ion current recorded by the mass spectrometer then arises only from the organic materials emerging from the g.c. column and the spectrometer itself acts as the g.c. detector. A disadvantage of using only a 20 eV ionizing energy is that the yield of ions is very much below that obtained at 70 eV and may give an unacceptably low sensitivity to the mass spectrometer.

The alternative to using a splitter is actual separation of the carrier gas from eluted organic material and allowing only the latter into the ion source. The simplest way of achieving this is to trap the organic part of the g.c. effluent first and then put the trapped material into the mass spectrometer as with an ordinary sample via one of the inlet systems. This method increases the amount of handling which must be done in any analysis and can scarcely be classed as a g.c.–m.s. combination.

A device interposed between the exit of the g.c. column and the inlet to the ion source which separates carrier gas from entrained organic material is called a *molecular separator*. There are four main types of molecular separator used with g.c.–m.s. combinations.

In the *Ryhage* separator (figure 5.10) carrier gas plus organic material exit from a small jet into a high vacuum. The carrier gas, helium, of lower molecular weight than organic compounds, diffuses faster from the stream of gas exiting from the jet. The gas stream is directed towards a small orifice and when it reaches this point much of the carrier gas plus some of the organic material has diffused from the stream and been pumped away. The residual organic material with any remaining carrier gas is passed either into a second similar separator stage to improve the degree of enrichment or it is passed straight into the ion source. Good enrichment is obtained with only a one-stage separator

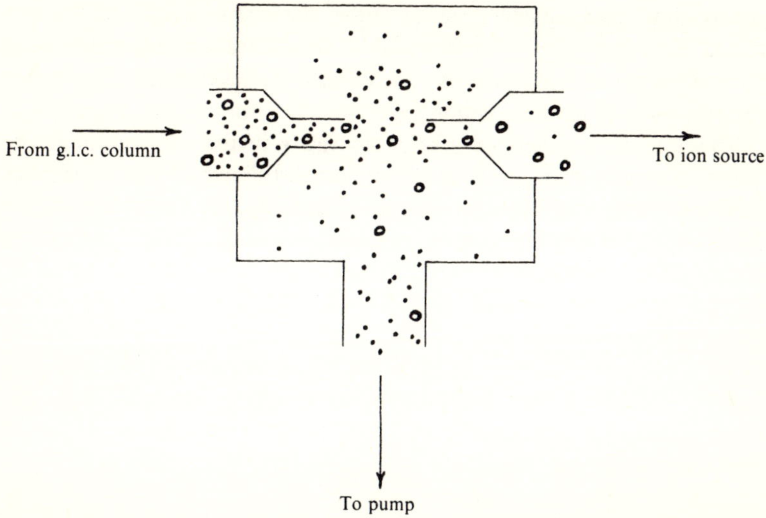

Figure 5.10. Ryhage separator showing enrichment of organic material (○) in carrier gas (•).

but much of the organic material is also pumped away so that the overall efficiency of transfer from the g.c. column into the ion source is not high.

The *Watson–Biemann* type of molecular separator (figure 5.11) uses the principle of differential molecular effusion through a porous partition to achieve enrichment of the organic phase entrained in the carrier gas. The low molecular weight helium effuses faster than organic

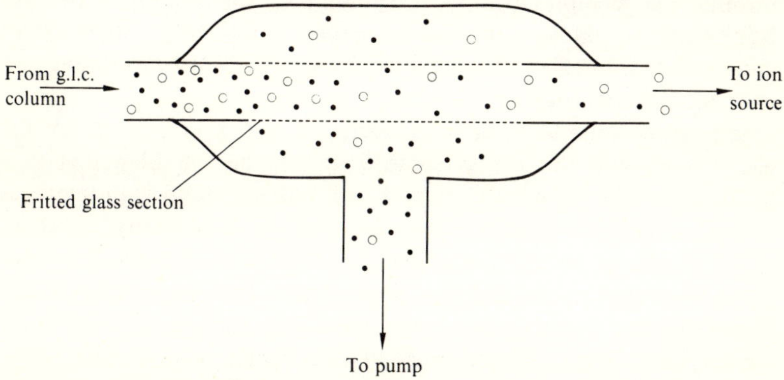

Figure 5.11. Watson–Biemann separator showing enrichment of organic material (○) in carrier gas (•).

compounds, particularly those of higher molecular weight, through a fritted glass partition. The material effusing through this partition is pumped away and the enriched g.c. effluent passes into the ion source; the efficiency of transfer is not high.

The *Llewellyn* separator works on the principle of solution with diffusion to give good transfer efficiencies. A thin membrane of silicone rubber is interposed between the g.c. effluent and the ion source (figure 5.12). At the membrane surface, organic molecules dissolve and diffuse

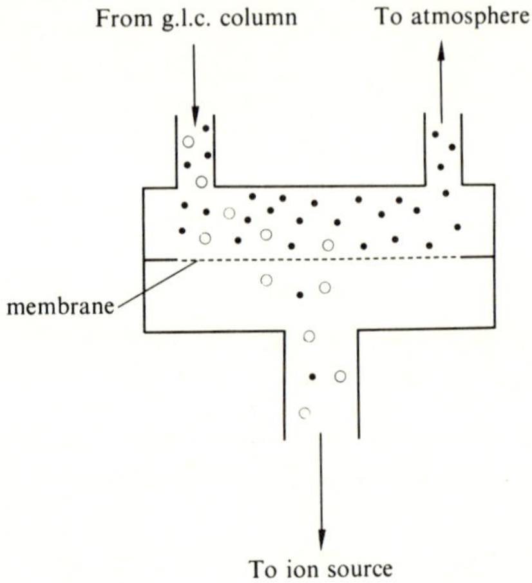

Figure 5.12. Llewellyn separator showing enrichment of organic material (○) in carrier gas (•).

through the membrane into the ion source. The helium carrier gas being scarcely soluble in the silicone permeates it to only a slight extent and most of it in effect 'bounces' off the surface of the membrane and does not diffuse through.

Separators must be capable of operating at the temperature of the g.c. column and this places some limitation on the Llewellyn type because the silicone membrane used will not withstand high temperatures. The silicone membrane may also give rise to 'memory' effects in which a residual amount of material from one peak of a gas chromatograph is caught up by the next peak eluted. Thus, although the gas chromatogram would indicate two completely separated compounds, the mass spectrum of the second component would be that of a mixture with some of the first component. These memory effects are not usually observed to any great extent with the other separators.

Preferably, there should be no metal surfaces in the separators to cause alterations in the organic material by metal catalysis.

Recently, a *variable* separator has been introduced which can be adjusted to accommodate flows of g.l.c. carrier gas from the low rates of capillary columns to the high rates of packed columns. A very narrow, adjustable gap along the length of the gas flow through the separator provides the degree of enrichment required.

By using a chemical-ionization source (§5.9), it is possible to dispense with a molecular separator. Methane can be used both as the carrier gas in gas chromatography and as the reactant gas in chemical ionization sources. Thus, the methane carrier gas from the g.c. column may be passed directly into the ion source where it then acts as the reactant (ionizing) gas and entrained organic compounds eluted from the column are 'chemi-ionized'.

Emergence of peaks from g.l.c. column. Unless arrangements can be made to control the flow of gas into the ion source, the time taken for a 'peak' of organic material from the g.c. column to pass into the source will vary. A g.c. peak may take only a few seconds to emerge from the column as in capillary work, or it may take tens of seconds as with packed columns. To be able to measure the mass spectra of compounds emerging from the g.c. column in a very short time it is necessary to be able to scan a mass spectrum in only a few seconds. At low resolution the mass spectrum can be scanned in a very short period of time and, by using a computer, such fast scans may be made at high resolution to yield accurate masses of all the ions.

The combination of a computer with a g.c.–m.s. set-up leads to a versatile piece of equipment. As peaks emerge from a g.c. column, they are scanned by the mass spectrometer, and the results stored on a magnetic disc. The mass spectra may then be recalled at will from the computer or the computer can compare the stored spectra with those in a library to provide an automatic identification of the material from the g.c. column (§4.1).

As a g.c. peak emerges from the column the amount of eluted organic material increases to a maximum and then decreases. Thus, the quantity of material passed into the ion source changes with time. During the scanning of the mass spectrum, the amount of material in the ion source is not constant and therefore, the intensities of the peaks in the mass spectrum will not match those in the mass spectrum of the material when maintained at a steady pressure in the source. To compensate for this, it is necessary to record continuously the total ion current in the mass spectrometer and then correct all the peak heights to allow for changes in this current. Figure 5.13 shows the effect of this correction on the appearance of a mass spectrum.

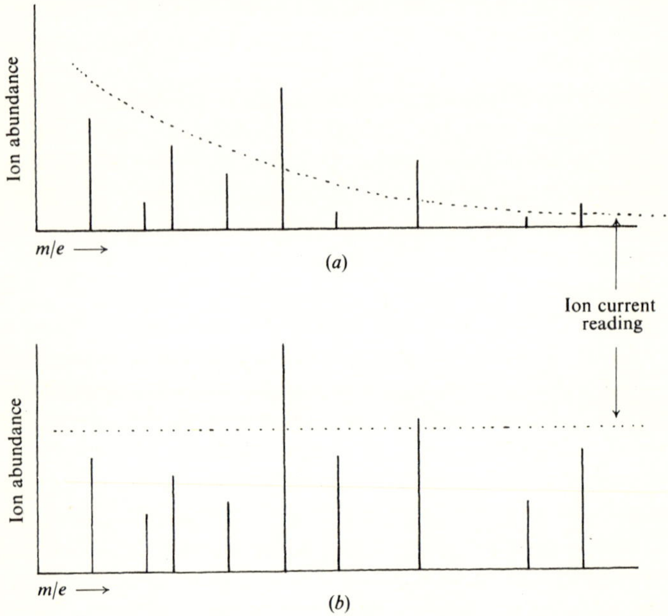

Figure 5.13. Changes in appearance of a mass spectrum with variation in total ion current during measurement. Spectrum (*a*) obtained with steadily falling ion current. Spectrum (*b*) is from same material as in (*a*) but obtained with a constant total ion current.

5.4. Ionization and appearance potentials. The least energy required to remove an electron from an atom or molecule is the ionization potential. Strictly, this should be the first ionization potential and correspond to the first adiabatic ionizing transition in a molecule. Morse curves for a simple diatomic molecule, M, are illustrated in figure 5.14. Except at high temperatures, most molecules will predominantly populate the zeroth vibrational state ($\nu = 0$) and this is the only vibrational state of the molecule considered here. A vertical transition such as v (figure 5.14) on ionization to the ground state ion, $M^{+\cdot}$, produces the ion in an excited vibrational state ($\nu = 3$); this transition is $0 \rightarrow 3$. The shaded area in figure 5.14 outlines the range in which

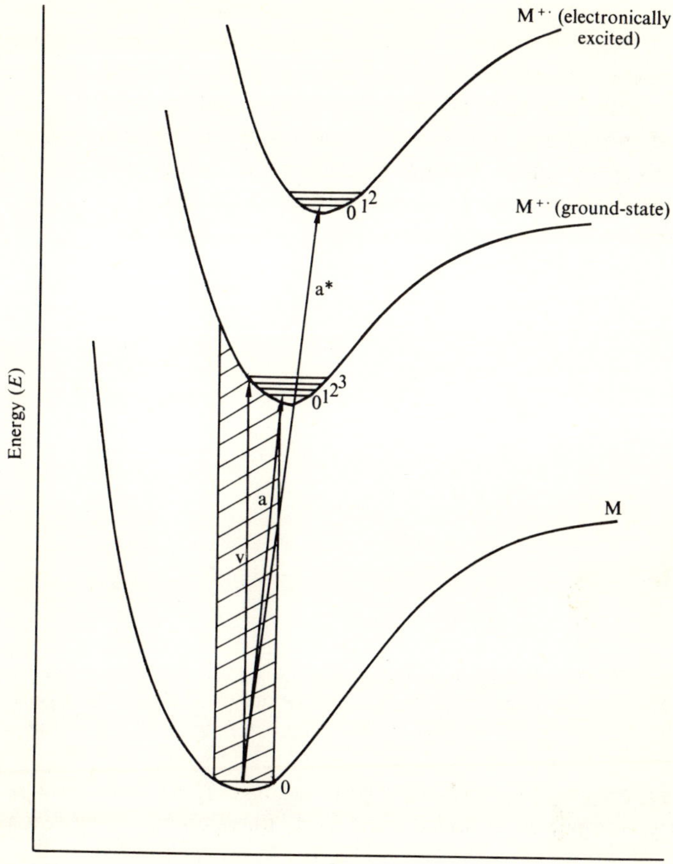

Figure 5.14. Morse curves showing ionizing transitions from molecule (M) to ground state and electronically excited ion ($M^{+\cdot}$). The transition (v) is a 'vertical' one, and transitions (a, a*) are adiabatic ones.

vertical transitions are possible from the $\nu = 0$ state of the molecule, M, to different vibrational states of the ion, $M^{+\cdot}$. One of these vertical transitions is called adiabatic (a) since ionization to the ground state ion, $M^{+\cdot}$, produces the ion in its lowest vibrational state ($\nu = 0$); this is a $0 \rightarrow 0$ transition. The $0 \rightarrow 0$ transition is that corresponding to the first ionization potential. The $0 \rightarrow 1$, $0 \rightarrow 2$, etc., transitions are closely separated in energy and are due to vibrational energy levels in the ground state of the ion; these transitions follow each other by a few tenths of an electronvolt. A $0 \rightarrow 0$ ionizing transition (a* in figure

5.14) to form the ion ($M^{+\cdot}$) in its first excited state (ionization from the second molecular orbital level) corresponds to the second ionization potential of the molecule (M). Similarly, there are third, fourth, and higher potentials each of which is accompanied by a series of closely spaced vibrational levels. Rotational energy is frequently neglected in mass spectrometry because the amount of energy involved is slight compared with the total excess of energy in the ion.

The first ionization potential (IP) of ethane is shown in figure 5.15,

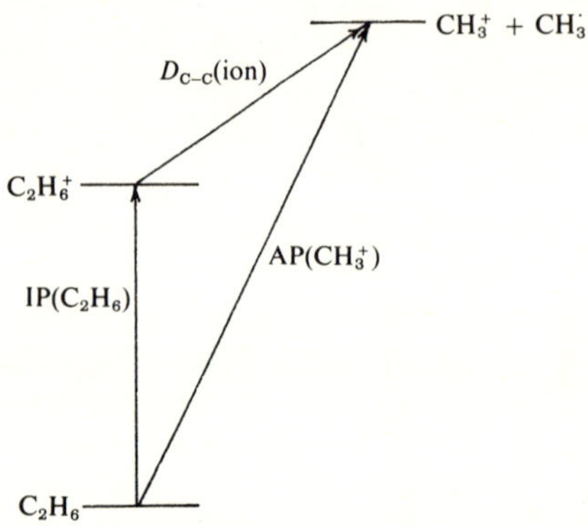

Figure 5.15. Energy diagram showing relationship between appearance potential of CH_3^+ from C_2H_6 and the ionization potential of C_2H_6. For a ground state reaction, the C—C bond energy in the ion (D_{C-C}) is $AP(CH_3^+) - IP(C_2H_6)$.

and represents the energy change in reaction (5.6). If gradually more energy is imparted in forming the C_2H_6 ion a point will be reached at

$$C_2H_6 \longrightarrow C_2H_6^{+\cdot} + e \tag{5.6}$$

$$C_2H_6^{+\cdot} \longrightarrow CH_3^+ + CH_3^{\cdot} \tag{5.7}$$

which the ion contains enough vibrational energy for it to fragment as in reaction (5.7). The energy at which the fragment ion, CH_3^+, first appears is the *appearance potential* for that ion (the appearance potential of a molecular ion is its first ionization potential). These ionization and appearance potentials can provide fundamental thermochemical information in chemistry, and mass spectrometers may be used to measure them. In the example of ethane given here, the C—C bond dissociation energy may be obtained from the appearance potential of the CH_3^+ ion and the ionization potential of the CH_3

radical. Equations (5.8) illustrate how the dissociation energy D_{C-C} is calculated, assuming there are no excess energy terms.

$$\begin{array}{llll} C_2H_6 & \longrightarrow CH_3^+ + CH_3^{\cdot} + e; & AP & \\ CH_3^{\cdot} & \longrightarrow CH_3^+ + e; & IP & (5.8) \\ \hline \therefore C_2H_6 & \longrightarrow CH_3^{\cdot} + CH_3^{\cdot}; & D_{C-C} = AP - IP & \end{array}$$

Since a mass spectrometer is a detector of ions it is only necessary to increase the energy applied to the molecules until molecular or selected fragment ions appear at the collector. The incident energy may be photons from a vacuum monochromator or electrons. A graph of the ion current versus the incidental radiation energy is an *ionization efficiency curve* as illustrated in figure 4.4 for electron-impact ionization. The onset of ionization is the ionization potential, but it can be difficult to assign an accurate value to it, because the position of onset may be somewhat indeterminate. The latter remark applies particularly to electron-impact derived values of ionization potentials and there seems little doubt that many of the older values determined this way will need to be corrected. More recently introduced methods of mathematical smoothing and analysis of raw data promise greater accuracy and consistency.

Appearance potentials determined by mass spectrometry may have dubious significance. In the example given above (5.8), it was assumed there were no excess energy terms or that these cancelled. These excess energy terms arise in several ways but probably the most important are (i) the molecular and electronic *structures* of the ions produced are *assumed* ones, (ii) the energy of activation for the reverse reaction is assumed to be zero i.e. AP = IP + D_{C-C} in (5.8) only if the activation energy for the reverse reaction is zero (see figure 4.19), (iii) for CH_3^+ ions to be collected in the mass spectrometer the $C_2H_6^+$ ion in (5.8) must decompose at a measurable rate and must therefore contain some excess of internal energy. If results determined by mass spectrometric procedures agree with those found by other methods it may be supposed that the assumptions are satisfactory. Where no check is available, the assumptions must be kept firmly in mind and results regarded as open to question. There are many mass spectrometric results which do not agree with thermochemical arguments based on other methods.

5.5 Isotope analysis and labelling. The measurement of isotope abundances by mass spectrometry can be made readily to a fair degree of accuracy. If relative abundances are required closer than about 3 per cent accuracy, special precautions must be taken and mass spectrometers have been designed specifically for isotope analysis. These spectrometers compare two beams of ions simultaneously and provide very accurate results by null methods. The incorporation or elimination of isotopic labels into organic compounds during a chemical reaction can be followed conveniently by mass spectrometry.

One of the simplest uses of isotope labelling is the determination of the number of active hydrogen atoms in a molecule. Treatment of a ketone, $R_2CH \cdot CO \cdot CH_2R$, having three enolizable hydrogens with either D_2O or CH_3OD containing a slight amount of alkali will give the deuterated ketone, $R_2CD \cdot CO \cdot CD_2R$, so that the molecular weight increases by three mass units. The increased molecular weight is readily detectable by mass spectrometry and it is unnecessary for there to be complete incorporation of label. Indeed, it may happen that even if incorporation of deuterium were complete, on putting the labelled compound into the mass spectrometer, exchange with protons from absorbed water on the surfaces inside the machine would occur; the incorporation of the isotopic label would then seem incomplete in the first place.

Labelling of organic compounds specifically for mass spectrometric work is usually done either to clarify a fragmentation sequence or to investigate the mechanisms of mass spectrometric reactions. The 'shift' technique can be used to help unravel fragmentation pathways. For example, the mass spectrum of a methyl ester may be compared with that of the trideuteriomethyl ester; all peaks which shift by three mass units must incorporate the methyl group.

Most labelling has been done to gain some insight into mass spectrometric fragmentation reactions and may be illustrated for the case of toluene.

The fragmentation of toluene in the mass spectrometer has been investigated extensively; the major features of the fragmentation are shown in scheme (5.9). The elimination of hydrogen from the molecular ion of toluene was at first believed to occur from the side-chain methyl group to give the benzyl cation, C_7H_7, at *m/e* 91. However, various deuterium labelling experiments showed that the ejected hydrogen was almost randomly derived from any of the original eight hydrogen atoms in the toluene molecule. Further, ^{13}C-labelling of toluene in the side-chain or in the ring showed that the ejection of C_2H_2 from the ion at *m/e* 91 was again a random process in that the labelled carbon atom

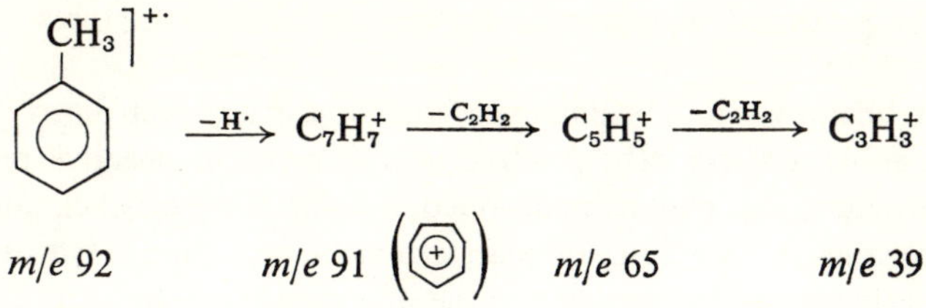

(1)

(5.9)

was not specifically retained in or ejected from the decomposing ion at *m/e* 91 (C_7H_7). These observations, together with measurements of the heat of formation of the ion, C_7H_7, suggested that a tropylium structure (1; 5.9) was plausible for the ion. If the toluene molecular ion at *m/e* 92 first rearranged to a hydrotropylium ion, the random loss of hydrogen to give the tropylium ion at *m/e* could be explained and also the random loss of ^{13}C-label as C_2H_2 from this ion was understandable (5.9).

More recent experiments with the doubly ^{13}C-labelled toluene (1; 5.10) have shown that, although the two ^{13}C-labels are adjacent in the molecule before ionization, when the ion at *m/e* 91 fragments, the loss of ^{13}C-label in the ejected C_2H_2 corresponds to complete randomization of all the carbon atoms as well as hydrogen in the ion at *m/e* 91

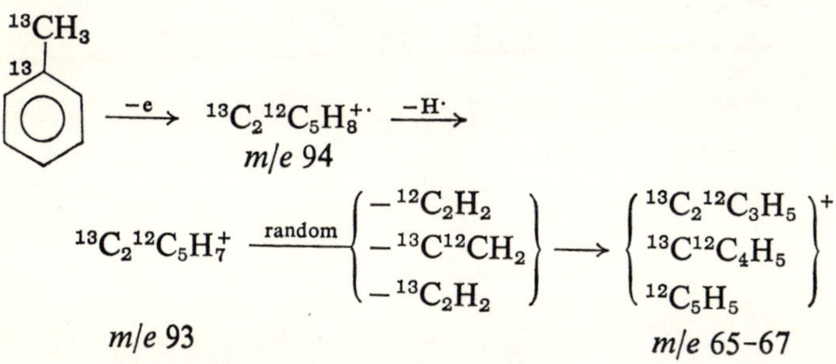

(5.10)

(5.10). The randomization of the carbon atoms may occur in the molecular ion which would not therefore have the toluene structure; the randomization could occur in the 'tropylium' ion at *m/e* 91 before it fragments; the randomization may be the result of the ion at *m/e* 91 not having a tropylium structure. Although the labelling experiments on toluene have produced some surprising results and given some insight into the fragmentation 'mechanism' of its decomposition, they have not produced unequivocal answers as to the structures of the ions involved.

To be unequivocal, labelling experiments in mass spectrometry need to give either 100 per cent positive or 100 per cent negative results. Intermediate values may be interpreted as scrambling of label, alternative mechanisms operating, or rearrangement occurring. Hydrogen is particularly prone to scrambling in the mass spectrometer, and deuterium labelling experiments are frequently negated by this effect.

Isotope labelling of organic compounds for mass spectrometry does not require the isotope to be radioactive. For good quantitative results and less ambiguity it is necessary to have as great an enrichment as possible of the isotopic label. The methods used to introduce labels are legion, but often some ingenuity is required in the synthesis of labelled compounds from expensive starting materials, the available range of which is limited. Labelled acetic acid, $CH_3{}^{13}CO_2H$, enriched to 65 per cent with ^{13}C is expensive but is in fact one of the cheapest ^{13}C-labelled compounds commercially available. By working on a small scale, the cost of the labelling experiment may be reduced considerably. The following syntheses of ^{13}C-labelled 1,4-diphenylbutadiene required a total of about 100 mg of acetic acid (5.11).

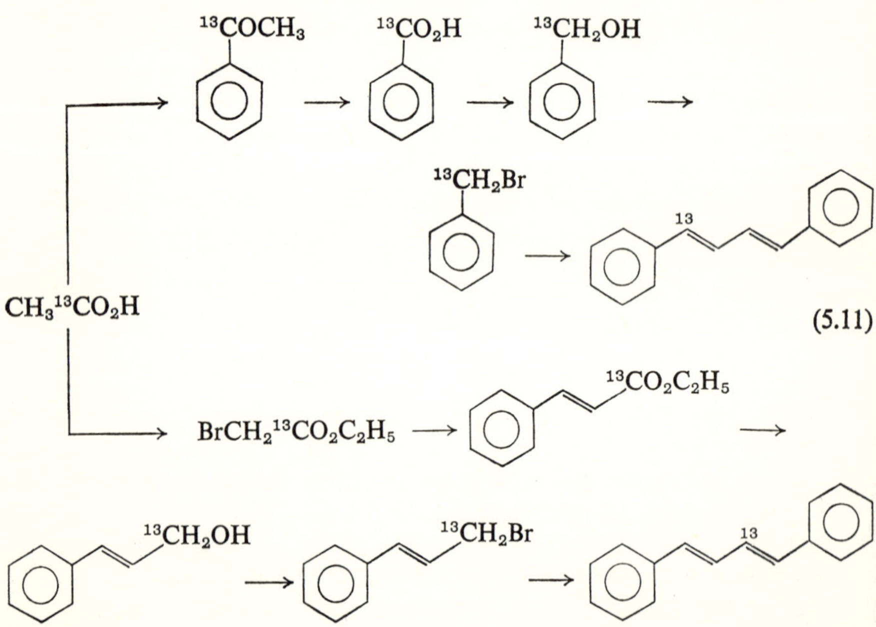

(5.11)

The incorporation of deuterium labels is usually easier and the required chemicals cheaper than for ^{13}C-labels. Thus, D_2O or CH_3OD may be used to exchange deuterium for hydrogen in amines, amides, alcohols, and enolizable carbonyl groups. Similarly, lithium aluminium deuteride is readily available and can be used to reduce many compounds. The use of an ^{18}O isotope is relatively expensive and fewer ^{18}O-labelled compounds are available compared with deuterium and ^{13}C.

Further examples of the use of isotopic labelling in mass spectrometry appear in §4.9 on rearrangements.

5.6. Analysis of mixtures. It may be suspected from examination of a mass spectrum that the sample under investigation is a mixture of compounds. Alternatively, it may be known that the sample is a mixture and mass spectrometric analysis is required.

Quantitative analysis of a mixture by mass spectrometry is difficult even when the components of the mixture are known. The components will generally volatilize into the mass spectrometer at different rates so that the appearance of the mass spectrum of a mixture is usually time dependent. Therefore, to obtain the proportions of the components of a mixture it is insufficient to record one mass spectrum; spectra must be recorded at intervals throughout the time any sample remains and an integration of the spectra must be carried out to determine the true proportions of components. If each component of a mixture is known to give a unique peak in the mass spectrum, the integration may be carried out after measuring the peak height at regular time intervals. Figure 5.16 shows how this integration is performed but to be quantitative the response of the mass spectrometer per unit of peak height must be predetermined. Even so the response of the mass spectrometer to each of the components is unlikely to be the same and a correction must be made for this factor too. Quantitative analysis of a mixture containing unknown components is unlikely to be very accurate and can be highly misleading.

Qualitative analysis of mixtures is a much simpler undertaking. If the individual mass spectra of the possible components of a mixture are available it is a simple task of inspection to determine which components are present. The only likely difficulties in these cases occur when the components have very similar mass spectra with many common ions, as for example with a mixture of isomeric alkanes. Often it is only necessary to find the molecular ion peak and a few of those of the major

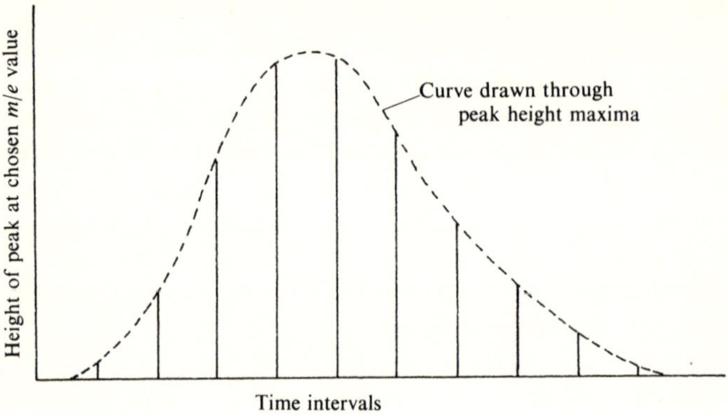

Figure 5.16. Diagram illustrating how area of curve drawn through peak height maxima obtained at regular time intervals is obtained to give an estimate of amount of a compound.

fragment ions to identify a component of a mixture when its mass spectrum is known.

The analysis of mixtures of totally unknown composition is more difficult. Although the mass spectrum of a component of a mixture may have been determined at some previous time, if its presence in the mixture is unsuspected, it will need careful analysis to dissect the individual spectrum from the peaks of the mixture spectrum. Consider a mixture of two components (A, B) giving a mass spectrum at some time t_1 (figure 5.17*a*); at some later time t_2, the spectrum of the mixture will have changed due to the differential volatilities of A, B (figure 5.17*b*). Figure 5.17*a*,*b* also shows the individual spectra of A, B at t_1, t_2 as well as the appearance of their mixture spectra at those times. The relative peak heights of ions in the spectrum of A or B individually will remain constant with time, assuming there are no changes in A or B chemically. Therefore, by measuring the relative heights of all the peaks in the spectra of the mixtures taken at times t_1, t_2, it is a straightforward mathematical exercise to identify which peaks belong to A and which to B. Thus, the individual spectra of A and B may be determined from the two spectra of their mixtures taken at some time apart and it is not necessary to know the spectra of A and B beforehand. A fairly simple computer program can be used to do the checking and sorting of the mixed spectra into the individual spectra of A and B. Computer programs have been written for dealing with mixtures of more than two components even when the components contain some common ions.

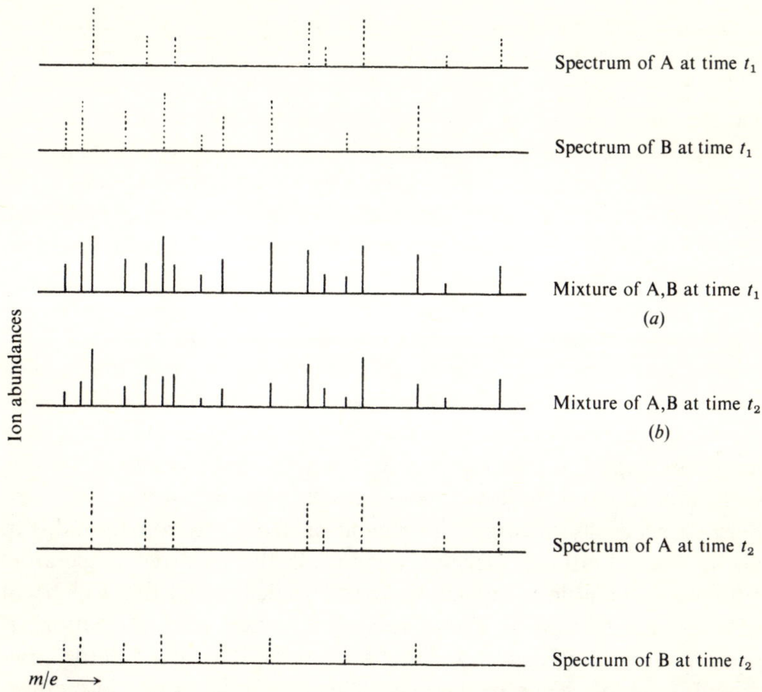

Figure 5.17. Appearance of mass spectrum of a mixture of two compounds (A, B) at (*a*) time t_1, (*b*) time t_2. For comparison, the diagram also shows how the individual spectra of A and B would appear at times t_1, t_2. Note how the *relative* ion abundances change in the mixture spectra but not in the individual spectra.

Other methods of examining the complexity of a mixture depend mostly on identifying the molecular ion of each component of the mixture. In this respect, field- and photo-ionization spectra are ideal because they normally contain intense peaks due to molecular ions with few fragment ions. Alternatively, when using an electron-impact source, the electron-beam energy can be reduced to about 10 eV when fragmentation is greatly reduced and molecular ions are prominent in the spectrum. One method of finding molecular ions depends upon deliberately increasing the pressure in the ion source to cause ion–molecule collisions. At the higher source pressures, hydrogen capture occurs and the intensity of the M + 1 ion increases. The technique is very similar to chemical-ionization which also leads to good molecular ions (see §5.9).

$$M^{+\cdot} + M \longrightarrow MH^+ + [M-H]^{\cdot}$$

Probably the easiest way of dealing with mixtures by mass spectrometry is to separate them into their components first! The combination of gas chromatography and mass spectrometry (§5.3) provides an excellent means of separation followed immediately by recording of mass spectra. It is also possible to check the purity of a peak emerging from a gas chromatographic column by recording a mass spectrum both at the beginning and at the end of the peak. Any differences in composition between the start and end of the gas chromatographic peak would be apparent from the mass spectra.

5.7. Sample volatility and directed fragmentation. The volatility of an organic compound is of considerable importance in obtaining a mass spectrum. At an ion source pressure of 10^{-6} torr many so-called non-volatile compounds have a high enough vapour pressure to give a good mass spectrum. For many organic compounds, some degree of heating is required to get a sufficiently high vapour pressure even at 10^{-6} torr. Because heat may cause rearrangement or decomposition and adds to the internal vibrational energy of a molecule, the lower the temperature required to volatilize the sample the less likelihood there will be of changes in structure or excessive fragmentation after ionization. If possible, it is preferable to convert involatile or slightly volatile compounds into volatile derivatives e.g. acids may be converted into esters. Similar problems occur in gas chromatography where volatility is important also, and similar solutions to those problems apply in mass spectrometry. In the latter case there is an added incentive to making derivatives other than considerations of volatility. It is often possible to increase the prominence of a molecular ion while at the same time increasing volatility. For example, a phenyl ester of a peptide gives a more intense molecular ion peak than the methyl ester and increases the volatility of the peptide almost as much. The effect of the derivative on the appearance of the mass spectrum must be considered along with its effect on volatility because introduction of a new grouping into a compound can lead to a large change in its spectrum. When derivatives are prepared specifically to induce a particular mode of ion decomposition it is known as *directed fragmentation.* Examples of this technique are the location of double bonds in alkenes (reaction 4.8) and the directed fragmentation of ketones by ketal formation (5.12). In the latter, the positions of substitution in the cyclic ketone can be determined from the masses of the fragment ions derived by cleavage of the ketal as shown (*m*/*e* 113, 99).

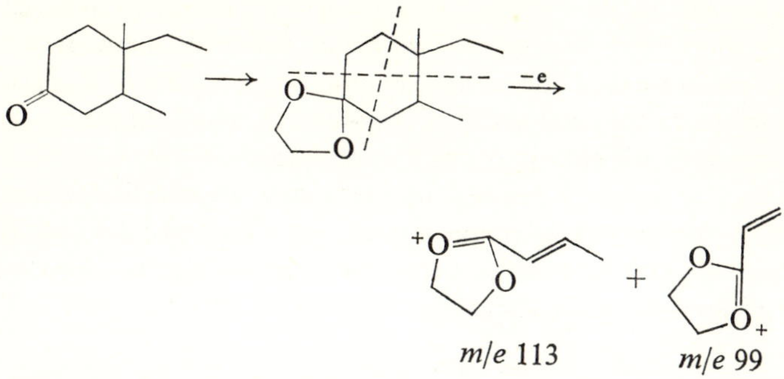

(5.12)

By increasing volatility, compounds of quite high molecular weight can be vaporized into the ion source. A decapeptide is unlikely to be sufficiently volatile to give a mass spectrum without considerable thermal decomposition. However, acetylating the terminal amino-group, methylating the terminal carboxyl group, and preventing hydrogen bonding by methylating all the amide bonds affords a derivative which is readily volatile even with a molecular weight of about 1500. The following methods of making derivatives are used.

Methylation. Acids and phenols with diazomethane yield esters and ethers respectively in a generally clean reaction. *N*-methylation of amides and *O*-methylation of alcohols can be effected with methyl iodide and silver oxide in dimethylformamide; there are now variations in this procedure, and sodium hydride for example may be used as the base. The *O*-methylation of polyols such as sugars is a good means of making them volatile. Conversion of ketones and aldehydes to ketals and acetals is useful not only for volatility reasons but also because the ketal or acetal grouping 'directs' the mass spectrometric fragmentation along known pathways and can give valuable structural information.

Silylation. Conversion of sugars and glycosides into trimethylsilyl derivatives is a clean, simple reaction frequently used to increase their volatility.

Acetylation. Alcohols, phenols, and amines are readily acetylated under mild conditions to give volatile derivatives. Often trifluoroacetylation is used instead because these derivatives are equally as volatile as acetyl derivatives and the products of reaction are easier to remove.

Reduction. Treatment of acids and amides with lithium aluminium hydride converts them quickly into more volatile alcohols and amines.

5.8. Ion cyclotron resonance spectroscopy. This method of mass spectroscopy has features which are different from the usual methods and merits some extra discussion. Two major differences between ion cyclotron resonance and the more commonly met methods of mass spectroscopy lie in the production of long-lived ions and the ability to examine ion–molecule collisions much more readily.

Under the influence of crossed magnetic and electric fields, ions describe a cycloidal path of frequency, $\omega_C = e.V/m$ and have a drift velocity, $v = V/H$ where V is the electric field, H the magnetic field, m the mass of the ion, and e its charge (figure 5.18). Both positive and

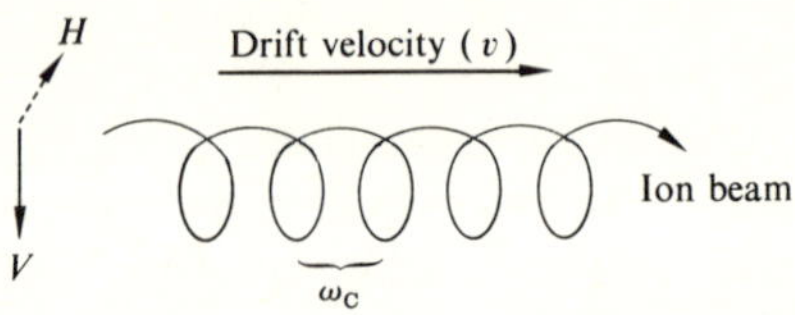

Figure 5.18. Side view of path of ion beam in ion cyclotron resonance spectroscopy (see figure 5.19 for top view). The ion cyclotron frequency (ω_C) and drift velocity (v) are shown. The electric field (V) and magnetic field (H) are crossed (V is in the plane of the paper; H is perpendicular to plane of paper).

negative ions obey the same laws and are distinguished only by the direction of rotation. In ion cyclotron resonance spectroscopy, ions are formed in an 'open' ion source (having no exit slits) and then subjected to the crossed electric and magnetic fields so that they travel through an analyser to the ion collector (figure 5.19). Space-charge effects would lead to the ions drifting to the walls of the analyser, and a 'trapping' potential is used to prevent this. In the collector region there is no

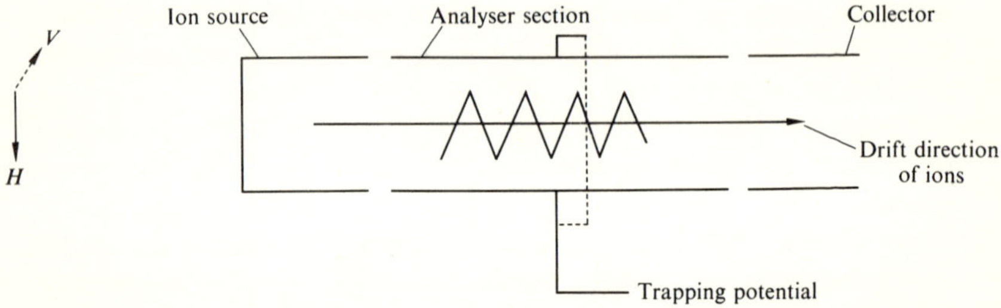

Figure 5.19. Top view of path of ion beam in ion cyclotron resonance spectroscopy showing side-to-side oscillatory motion of ions impressed on their cycloidal path (see figure 5.18). The ions produced in the source pass through an analyser section to a collector. A trapping potential prevents the ions striking the sides of the analyser. The electric field (V) is perpendicular to, and the magnetic field (H) is in, the plane of the paper.

trapping potential and the ions striking the walls there are recorded as the total ion current. Because the ions follow a long cycloidal path, the actual time taken to traverse the analyser is 5–10 milliseconds although the analyser itself is only a few centimetres long. The time may be compared with the microsecond time of flight in many other mass spectrometers. The long flight time in ion cyclotron spectroscopy leads to a high probability that ion–molecule collisions will occur before the ions are collected and it is possible therefore to investigate ion–molecule reactions.

Since all ions proceed through the analyser at the same drift velocity, examination of individual ion species is carried out by applying a radio-frequency electromagnetic field to the ion beam in the analyser. Although all ions have the same drift velocity, their individual cyclotron frequencies (ω_C) are proportional to their *m/e* values. When the frequency of the applied radio-frequency field equals that of the ion cyclotron frequency, absorption of energy occurs from the radio-frequency field; this resonance condition of energy absorption is measured to assign *m/e* values.

It is outside the scope of this book to discuss ion–molecule reactions (Knewstubb, 1969) but it may be mentioned that, apart from their intrinsic interest, they can give valuable thermochemical data such as proton and electron affinities. For organic mass spectrometry, the investigation of ion–molecule reactions by ion cyclotron resonance spectroscopy can yield considerable information on ion structures. The use of the method in this way may be illustrated by a single example. The ion resulting from six-centre rearrangement of 2-hexanone is considered to be the enol of acetone (5.13; 1); the ion resulting from simple fragmentation of the methylcyclobutanol (2) is also thought to possess the structure (1). These fragment ions collided with molecules in the analyser section of the ion cyclotron resonance spectrometer and subsequent ion–molecule reactions were investigated. No difference was found in the behaviour of the two ions (1) at *m/e* 58 from 2-hexanone or the methyl cyclobutanol (2), from which it was concluded the ions had the same structure. One typical ion–molecule reaction of the ion at *m/e* 58 is shown (5.14). The molecular ion at *m/e* 86 of the cyclobutanol (2) fragments to give the ion (1) at *m/e* 58 (C_3H_6O) which, by ion–molecule collision with a neutral molecule of the cyclobutanol (2), then affords an ion, $C_6H_{10}O$, at *m/e* 98. To differentiate the ion, C_3H_6O, at *m/e* 58 produced by the cyclobutanol (2) from the ion of the same composition produced by fragmentation of 2-hexanone,

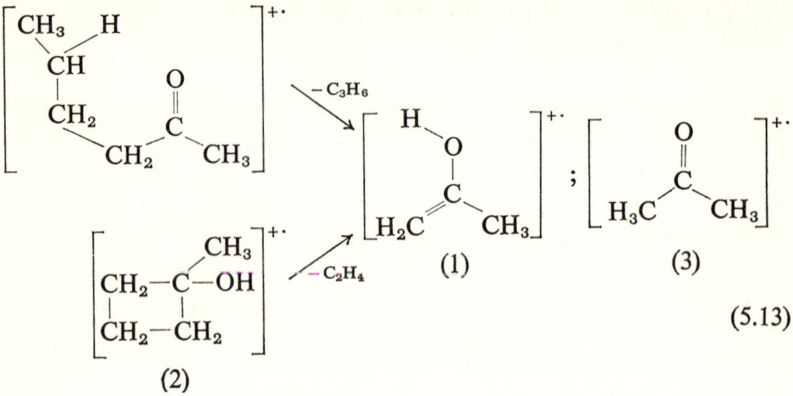

(5.13)

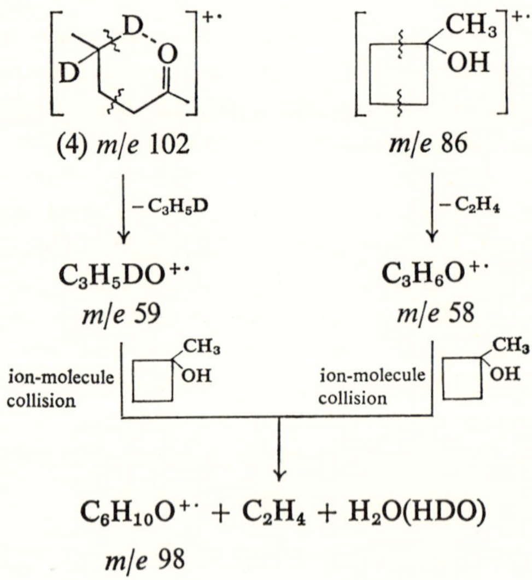

(5.14)

the latter was labelled with deuterium in the 5-position. The 5,5-d_2-hexanone (4) fragmented like 2-hexanone to give the ion (1) but now of composition C_3H_5DO at *m/e* 59. In the ion cyclotron resonance spectrometer, this ion at *m/e* 59 also underwent an ion–molecule reaction with neutral methylcyclobutanol (2) to give the ion, $C_6H_{10}O$, at *m/e* 98 (5.14). Thus, the ions, C_3H_6O, from methylcyclobutanol and C_3H_5DO from the hexanone (4) when produced together in the spectrometer, were both found to react with neutral methylcyclobutanol molecules to give the ion, $C_6H_{10}O$, at *m/e* 98. The molecular ion of acetone (5.13; 3) was investigated in the same way but its ion–molecule reactions were found to be different from those of the fragment ion (1). It may be concluded that the fragment ions of composition C_3H_6O derived from both the methylcyclobutanone (2) and 2-hexanone (5.13) have the same structure which is different from the structure of the molecular ion of acetone (3). Although structures of ions may be compared in this way by ion cyclotron spectroscopy, it should be borne in mind that the actual structures may not be as postulated since rearrangement may occur in an ion–molecule reaction.

Negative ion spectra may be measured as readily as positive ones by ion cyclotron resonance spectroscopy. An interesting application of this facility is the measurement of electronic excitations in molecules resulting from impact with low energy electrons. In the reaction sequence (5.15), electrons of energy e_1 are captured by SF_6 and the abundance of SF_6^- ions determined by ion cyclotron resonance. In the presence of other gas molecules

$$\begin{array}{ccccl} M + e_1 & \longrightarrow & M^* + e_2 & & (e_2 < e_1) \\ \begin{pmatrix}\text{neutral} \\ \text{molecule}\end{pmatrix} & & \begin{pmatrix}\text{electronically} \\ \text{excited}\end{pmatrix} & & \\ SF_6 + e_1(e_2) & \longrightarrow & SF_6^- & & \end{array} \tag{5.15}$$

energy may be transferred from the electrons with energy e_1 due to electronic excitation of the molecules; in this process the electrons after impact have a lower energy (e_2). The electronic excitation energy of the molecule is then $e_1 - e_2$. The change in the impacting electron energy leads to a change in the abundance of SF_6^- ions. Thus, by monitoring the ion current due to SF_6^- ions as the energy of the impacting electrons is gradually increased, electronic excitation processes in the other gas molecules can be investigated.

It is also possible to carry out a technique of double resonance spectroscopy by irradiating one ion at its ion cyclotron frequency whilst observing the effect of this on the abundance of another ion at a different frequency. If an ion–molecule reaction (5.16) is suspected, ion A^+ can be irradiated in the ion source at its cyclotron frequency, ω_A, and ion

$$A^+ + B \longrightarrow C^+ + D \tag{5.16}$$

C^+ observed at its resonance frequency ω_C. If C^+ ions are derived from A^+ ions, then changing the translational energy of A^+ ions by irradiation at ω_A changes the intensity of the peak due to C^+ and the interrelationship between A^+ and C^+ demonstrated. For example, the ion at *m/e* 59 resulting from a double McLafferty rearrangement in the molecular ion of 4-nonanone-1,1,1-d_3 has been considered to have either structure (1) or (2) in reaction scheme (5.17). The molecular ion of a ketone such as 4-nonanone is readily protonated by collision

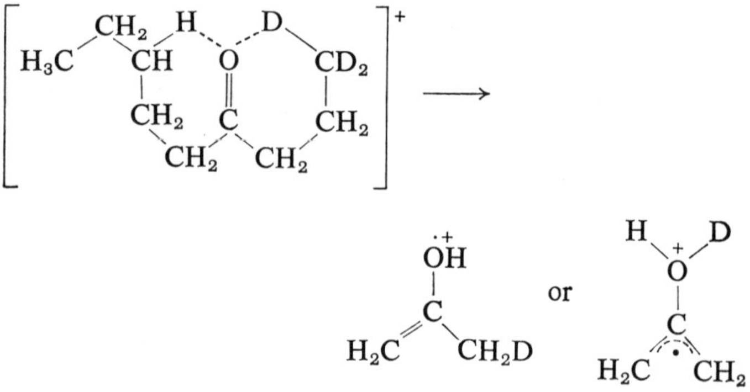

(5.17)

(1) (2)

$[C_3H_5D]$ *m/e* 59

with radical–ions having hydrogen on an oxygen atom (5.18). Therefore, collision of the ion at *m/e* 59 with 4-nonanone-1,1,1-d_3 (molecular weight 145) could lead to transfer of H^+ or D^+ with formation of protonated 4-nonanone at *m/e* 146 ($M + H^+$) or *m/e* 147 ($M + D^+$) depending on the structure of the ion at *m/e* 59 (1 or 2 in reaction 5.17). The ion at *m/e* 147 was observed at its cyclotron frequency and the ion at *m/e* 59 irradiated at its cyclotron frequency. No change was found in the abundance of the ion at *m/e* 147 and therefore it could not have

$$CH_3(CH_2)_4\overset{\overset{O}{\parallel}}{C}(CH_2)_2CD_3 + \underset{m/e\ 59}{C_3H_5D^{+\cdot}} \longrightarrow$$

$$\underset{\substack{[M + H^+]\ m/e\ 146 \\ [M + D^+]\ m/e\ 147}}{CH_3(CH_2)_4\overset{\overset{\overset{+}{O}\text{—}H(D)}{\parallel}}{C}(CH_2)_2CD_3} + C_3H_4D(H) \quad (5.18)$$

been formed by collision with the ion at *m/e* 59. On the other hand, the abundance of the ion at *m/e* 146 was strongly affected when the ion at *m/e* 59 was irradiated. It was concluded from this *double resonance* experiment that the ion at *m/e* 59 could not have had the structure (2) in (5.17) because transfer of deuterium from the ion to 4-nonanone was not detectable.

As well as recording a mass spectrum by measuring the energy loss of the irradiating field at each ion cyclotron resonance frequency, a mass spectrum can be recorded by monitoring the total ion current at high field strengths. Under these conditions, when resonance occurs, an ion is ejected onto the sides of the cell before reaching the collector and therefore there is a decrease in the total ion current. These changes in the total ion current at each *m/e* value yield the mass spectrum.

It is possible to assess the relative contributions of different ion–molecule reactions leading to a common ion. Consider the two ion–molecule reactions shown (5.19). If the mass spectrum could be recorded

$$\begin{aligned} A^+ + B &\longrightarrow C^+ + D \\ X^+ + B &\longrightarrow C^+ + Z \end{aligned} \qquad (5.19)$$

without ion A^+ (or X^+), the relative importance of the two reactions leading to the formation of C^+ could be assessed. In ion cyclotron resonance spectroscopy this assessment can be made by 'ejecting' A^+ (or X^+) from the cell by application of an oscillating electric field to the trapping plates (figure 5.19). Since the ions are oscillating between the trapping plates as well as following a cycloidal path, then at resonance the oscillations increase and the ions are ejected to the walls. Ejection of A^+ would reduce the abundance of C^+ ions to that due solely to their formation from X^+ ions. The technique is usable with any number of reactant ions so long as they can be uniquely ejected.

5.9. Chemical-ionization. The ion sources used for chemical-ionization operate at pressures of about 10^{-4} torr compared with 10^{-6} to 10^{-5} torr for electron-impact sources. At the higher pressure of 10^{-4} torr, the mean free path of a molecule or ion is about 2 mm; at 10^{-6} torr it is 200 mm. Therefore, after formation at a pressure of 10^{-4} torr, ions can collide with other neutral molecules many times before being drawn out of the ion chamber. Through these collisions, energy is equilibrated between the ions and molecules, so that in chemical-ionization the primary ions are generally considered to possess thermal energies i.e. energies corresponding to equilibrated ground state species at the source temperature. The primary ions are formed by electron-impact and on collision with neutral molecules a number of ion–molecule reactions may occur. For structural organic mass spectrometry, the most useful reaction of the primary ions is the formation of stable, even-electron secondary ions. Hydrogen, methane, propane, and isobutane have been employed as the *reactant gases* in this way, the principal secondary ions for the hydrocarbons being shown in table 5.2.

TABLE 5.2 *Principal secondary ions in reactant gases*

Reactant gas	Reactant ions produced
Methane	CH_5^+ (44%); $C_2H_5^+$(30%)
Propane	$C_2H_5^+$(15%); $C_3H_7^+$(60%)
Isobutane	$C_3H_7^+$(12%); $C_4H_9^+$(64%)

It is the thermally equilibrated, stable secondary ions which are the reactant species and which react with any other organic molecules present (5.20). Normally, to investigate the mass spectrum of an organic

$$R + e \longrightarrow [R^{+\cdot}]^* \xrightarrow[\text{mechanism}]{\text{collisional}} RH^+$$

[R⁺·]*: Excited primary reactant gas ion; RH⁺: Equilibrated secondary reactant gas ion.

$$RH^+ + M \longrightarrow [MH]^+ + [R]$$

$$RH^+ + M \longrightarrow [M-H]^+ + [RH_2] \qquad (5.20)$$

RH⁺: Reactant gas ion; M: Molecule under investigation; [MH]⁺, [M−H]⁺: Quasi-molecular ions; [R], [RH₂]: Neutrals

compound about 0.1 per cent of it is incorporated in the reactant gas. In such low proportion very few primary ions are formed by electron-impact with the organic molecule being investigated, but the stable secondary reactant gas ions can collide with the organic compound to produce tertiary ions; these latter are the ions of interest to the organic mass spectroscopist. The secondary ions are strong Lewis acids and during the formation of the tertiary ions, either a proton is donated by the reactant gas or a hydride ion is abstracted. The chemical-ionization of acetone in methane can be represented by equation (5.21) in which a secondary CH_5^+ ion donates a proton to acetone. The oxygen atom is considered protonated in the tertiary ion which has a molecular weight

$$CH_5^+ + CH_3\underset{\underset{\displaystyle O}{\|}}{C}CH_3 \longrightarrow CH_4 + \underset{\text{Quasi-molecular ion}}{CH_3\underset{\underset{\displaystyle {}^+O\text{—}H}{\|}}{C}CH_3} \tag{5.21}$$

greater by one than that of acetone; for this reason the tertiary ion is referred to as a quasi-molecular ion (QM^+).

If the quasi-molecular ions have sufficient energy, they dissociate to give even-electron fragment ions. Results obtained so far by chemical-ionization methods indicate that the quasi-molecular ions do not seem to possess large amounts of excess internal energy so that prominent quasi-molecular ions are observed in the mass spectrum together with a greatly reduced number of fragment ions when compared with electron-impact methods. Thus, many biologically important compounds such as terpenes, steroids, sugars, nucleosides, peptides, and alkaloids yield intense peaks in their mass spectra corresponding to quasi-molecular ions $(M + H)^+$.

The fragmentation patterns of chemical-ionization mass spectra appear to be easier to interpret than electron-impact spectra since they give fewer but more diagnostic fragment ions. The loss of water from ketones is commonly found in mass spectrometry but the chemical-ionization mass spectra of steroidal ketones exhibit marked selectivity in this reaction. The loss of water from the quasi-molecular ions of 5α-3-ketosteroids is not prominent but is most marked in the corresponding 5β-3-ketosteroids.

The actual degree of fragmentation may be altered by changing the reactant gas. Chemical-ionization of acetophenone gives $(M + H)^+$ quasi-molecular ions using either methane, propane, or isobutane as the reactant gas. However, the degree of fragmentation varied; with isobutane only quasi-molecular ions were observed; with propane QM^+ and $C_6H_5CO^+$ ions were found; with methane QM^+, $C_6H_5CO^+$,

$$\underset{QM^+}{[C_6H_5COCH_3 + H]^+} \longrightarrow \underset{\text{Fragment ions}}{C_6H_5CO^+,\ C_6H_5^+,\ CH_3CO^+}$$

$C_6H_5^+$, and CH_3CO^+ ions were formed. Thus, the mass spectrum can be changed by changing the secondary reactant ions.

With its prominent quasi-molecular ions and reduced fragmentation, chemical-ionization mass spectrometry is likely to find increasing use for structural work in organic chemistry. Unlike electron-impact-ionization, chemical-ionization gives mass spectra which reflect the fine differences existing between structural isomers.

If the reactant gas ions are formed from monatomic species there are no vibrational degrees of freedom and the ions carry specific amounts of energy. Argon, ionization potential 15.755 eV, has just that excess of energy after ionization and there is no possibility of its internal conversion to vibrational energy. In a charge-exchange reaction between reactant argon ions and organic molecules, the whole of the excess of energy (15.755 eV) is transferred to the organic molecule which is therefore ionized and given a large excess of energy. The organic molecule does have vibrational degrees of freedom and the excess of

$$A^{+\cdot} + M \longrightarrow A + M^{+\cdot}$$

energy present causes rapid fragmentation. The behaviour is then very similar to the fragmentation of odd-electron molecular ions produced by electron-impact, and is unlike the chemical-ionization discussed above in which thermally equilibrated reactant gas ions transfer smaller amounts of energy to form even-electron quasi-molecular ions exhibiting greatly reduced fragmentation. The interest in charge-exchange reactions with monatomic reactant gas ions lies in the knowledge of the precise amount of excess of energy given to the organic molecular ion. If the ionization potential of the organic molecule is 10.0 eV and argon is used as the reactant gas, then $15.755 - 10.0 = 5.755$ eV excess of energy is imparted to the molecular ion. Such an amount of energy appearing in most of the commoner types of organic molecular ions leads to rapid fragmentation.

5.10. Field-ionization. It is appropriate at this stage to collate the earlier scattered references in this book to field-ionization mass spectrometry.

Molecules approaching or adsorbed on a surface of high curvature maintained at a high positive potential are subjected to potential gradients of about 10^7 to 10^8 volts/cm. Under the influence of these fields, quantum tunnelling of a valence electron from the molecule to the anode takes place in about 10^{-12} seconds to give an ion–radical.

The ion–radical being positively charged is accelerated from the anode and collected in the normal way.

If ionization occurs at the electrode surface, the ion is subjected to the full electric field and is collected as a normal narrow peak in the mass spectrum. Ionization or fragmentation close to, but not at, the surface gives an ion with less than the full translational energy; such ions appear as 'tails' on the low mass side of normal peaks and are called 'fast' metastables. If an ion fragments between ionization and collection it appears as a normal metastable ion (§5.1.).

Field-ionization mass spectra are generally characterized by the low degree of fragmentation observed compared with electron-impact spectra. The formation of the molecular ion by tunnelling is thought not to produce an ion in a highly excited state and therefore with not enough excess of energy for extensive fragmentation.

Rearrangement processes have lower reaction frequencies (ν in the QET equations, §3.3) than simple bond cleavage reactions. It has been suggested therefore that, because of the short time the ions spend in the ion source compared with that in an electron impact source, there is little time for rearrangement and few such processes should be observed. However, other work suggests that rearrangement can take place in the strong electric fields, partly at least, *before* ionization. For example, 2-alkanones yield abundant normal ions at *m/e* 58 resulting from McLafferty rearrangement in field-ionization spectra. In some cases, the rearrangement ion is so abundant it forms the base peak in the mass spectrum.

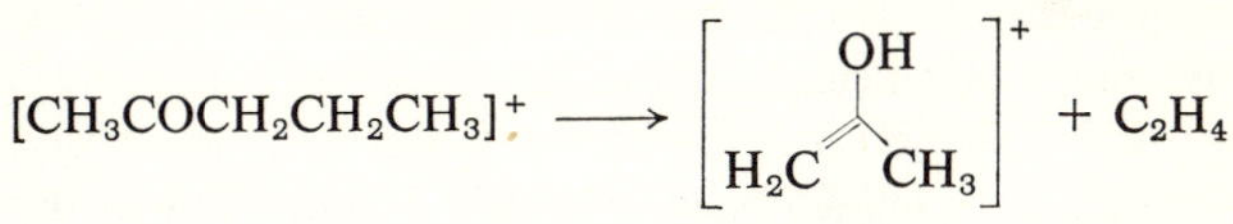

m/e 58

In field-ionization, fragmentation can often be explained as cleavage of a molecular ion which is strongly polarized in the electric field. Fragmentation may also occur through collision in the condensed phase around the electrode where energetic ion–molecule reactions can occur.

The metastable and normal ions in a field-ionization mass spectrum are very useful in structural organic chemistry because isomers can be distinguished; this is very often not the case in electron-impact work.

A modified field-ionization process has been used for thermally labile compounds. Normally, an organic compound is evaporated from a small oven near the positive electrode and molecules are ionized as they near or strike the surface. In the *field-desorption* method, the organic compound is put directly onto the electrode as a thin layer. No oven is required for vaporizing the sample, thereby reducing greatly any possible thermal effects. Abundant molecular, M, or M + H ions can be obtained from substances which otherwise give poor ones. For example, the field desorption spectrum of D-glucose shows an abundant M + H ion at *m/e* 181 with almost no fragment ions. The usual field-ionization spectrum where evaporation of the glucose onto the electrode is used, also has an abundant M + H ion (base peak) but shows several abundant fragment ions. The electron-impact spectrum of the compound has no M or M + H ions and only fragment ions up to *m/e* 73. The method promises to be of value for thermally labile compounds and those requiring such high temperatures to volatilize them that extensive decomposition occurs.

5.11. Relative abundances of isotope peaks. A mass spectrometer measures mass-to-charge ratios and the effect of elemental isotopes on the appearance of a mass spectrum has been described in §1.3. Ions containing carbon (^{12}C) are always accompanied by ions one mass unit greater because of the ^{13}C isotope. The natural abundance of ^{13}C is quite low so that it does not complicate the mass spectrum too much except for relatively large numbers of carbon atoms. Other elements of interest in organic chemistry such as nitrogen and oxygen also contain only low percentages of isotopes so that their abundances are of not much significance unless there are large numbers of these atoms in the ion. However, some elements have abundant isotopes as with chlorine and bromine (listed in table 1.3) and many of the metals used organometallic chemistry. When elements with abundant isotopes are found in mass spectrometry, it is necessary to have some idea of how the appearance of the mass spectrum will depend on their number i.e. to know what the *isotope pattern* will be.

Any ion (M) containing one chlorine atom will be accompanied by an ion at M + 2 of approximately one-third the abundance because the natural relative abundances of ^{35}Cl and ^{37}Cl are in the ratio 3:1. When there are two chlorine atoms in the ion (RCl_2), the calculation of the relative abundances of the isotope peaks (isotope pattern) becomes more complicated. Three ions (M, M + 2, M + 4) occur corresponding to the possible combinations of ^{35}Cl, ^{37}Cl (5.22). Simple calculation of probabilities shows that the ions M, M + 2,

$$\underbrace{R\,^{35}Cl\,^{35}Cl}_{M} \qquad \underbrace{R\,^{35}Cl\,^{37}Cl}_{M+2} \qquad \underbrace{R\,^{37}Cl\,^{37}Cl}_{M+4} \tag{5.22}$$

M + 4 will have relative abundances of 9:6:1 (figure 1.6). For more than two chlorine atoms, the calculation becomes more laborious, but expansion of the simple formula (5.23) yields immediately the number and relative abundances of the isotope peaks. In the formula (5.23), n is the number of atoms to be considered and a, b are relative abundances of the isotopes. For two chlorine atoms in an ion, $n = 2$, $a = 3$,

$$(a + b)^n \tag{5.23}$$

$b = 1$ and hence $(a + b)^n = (3 + 1)^2 = 9 + 6 + 1$, and these terms are the relative abundances of the chlorine isotope peaks in an ion containing two chlorine atoms. For three bromine atoms, $n = 3$, $a = 1$, $b = 1$ (^{79}Br:$^{81}Br \approx 1:1$, see table 1.3) and so $(a + b)^n = (1 + 1)^3 = 1 + 3 + 3 + 1$. Thus, an ion with three bromine atoms would give peaks in the mass spectrum at M, M + 2, M + 4, M + 6 in the relative abundances of 1:3:3:1. The reader familiar with nuclear magnetic resonance spectroscopy will note how the same formula (5.23) is used to obtain spin 'splitting patterns'.

When an ion contains two or more different elements with isotopes, the calculation *ab initio* of relative abundances becomes laborious and time-consuming. Fortunately, there is a simple short cut for obtaining the isotope pattern in these cases which is best exemplified by two specimen calculations (5.24, 5.25).

Firstly, an ion with two chlorine and two bromine atoms.

$$\left.\begin{array}{ll} \text{For } Cl_2, & (a + b)^n = (3 + 1)^2 = 9\ 6\ 1 \\ \text{For } Br_2, & (a + b)^n = (1 + 1)^2 = 1\ 2\ 1 \end{array}\right\}\text{relative abundances} \tag{5.24}$$

These relative abundances are combined,

$$\begin{array}{lrl} (9\ 6\ 1)(1\ 2\ 1) \therefore & (9\ 6\ 1) \times 1 = & 9\ \ 6\ \ 1 \\ & (9\ 6\ 1) \times 2 = & \quad 18\ 12\ 2 \\ & (9\ 6\ 1) \times 1 = & \qquad\quad 9\ 6\ 1 \\ \hline & \text{Total} & 9\ 24\ 22\ 8\ 1 \end{array}$$

that is, each of the terms relating to chlorine abundances (9 6 1) is multiplied by each term relating to the bromine abundances (1 2 1).

After each multiplication, the resulting terms are moved one column to the right and finally totalled. Thus, the isotope pattern for Cl_2Br_2 is five ions M, M + 2, M + 4, M + 6, M + 8 in relative abundances of 9:24:22:8:1.

Secondly, an ion having two sulphur atoms and one chlorine,

$$\begin{aligned} &\text{For } S_2, \quad (a+b)^n = (95+5)^2 = 9025\ 950\ 25 \\ &(^{32}S{:}^{34}S \approx 95{:}5, \text{ see table 1.3}) \\ &\text{For Cl}, \quad (a+b)^n = (3+1)^1 = 3\ 1. \end{aligned} \tag{5.25}$$

Combining these relative abundances (5.25) gives,

$$\begin{array}{rrrrr} (9025\ 950\ 25)(3\ 1) \therefore (9025\ 950\ 25) \times 3 = & 27075 & 2850 & 75 & \\ (9025\ 950\ 25) \times 1 = & & 9025 & 950 & 25 \\ \hline \text{Total} & 27075 & 11875 & 1025 & 25 \\ \hline \text{or} & 1083 & 475 & 41 & 1 \end{array}$$

The isotope pattern consists therefore of four peaks M, M + 2, M + 4, M + 6 in the ratios 1083:475:41:1. The M + 6 peak is very much less abundant than the other three and might be ignored in the mass spectrum. Notice that the ratio (2.3:1) of the M and M + 2 peaks is quite different from the 3:1 it would be if only chlorine and no sulphur were present. Although the natural abundance of ^{34}S is quite modest, its contribution to the spectrum would not be missed by an experienced mass spectrometrist.

On that note, emphasizing the value of experience in the application of mass spectrometry to structural problems in organic chemistry, this introductory text must be closed.

Bibliography

Advances in Mass Spectrometry, **1–4**, Institute of Petroleum, London.

Barber, M., and Elliott, R. M. (1964). 12th ASTM E-14 Meeting on Mass Spectrometry, Montreal.

Bentley, T. W., and Johnstone, R. A. W. (1970). *Advances in Physical Organic Chemistry*. **8**, Ed. V. Gold, Wiley, New York.

Beynon, J. H. (1960). *Mass Spectrometry and Its Applications to Organic Chemistry*, Elsevier, Amsterdam.

Beynon, J. H., and Williams, A. E. (1963). *Mass and Abundance Tables for Use in Mass Spectrometry*, Elsevier, Amsterdam.

Beynon, J. H., Saunders, R. A., and Williams, A. E. (1965). *Z. Naturforschung*, **20a**, 180.

Beynon, J. H., Saunders, R. A., and Williams, A. E. (1968). *The Mass Spectra of Organic Molecules*, Elsevier, Amsterdam.

Beynon, J. H., Caprioli, R. M., Baitinger, W. E., and Amy, J. W. (1970). *Org. Mass Spectr.* **3**, 455.

Biemann, K. (1962). *Mass Spectrometry, Organic Chemical Applications*, McGraw-Hill, New York.

Brown, P., and Djerassi, C. (1967). *Angew. Chem.* (Int. Ed. Engl.), **6**, 477.

Budzikiewicz, H., Djerassi, C., and Williams, D. H. (1964*a*). *Interpretation of Mass Spectra of Organic Compounds*, Holden-Day, San Francisco.

Budzikiewicz, H., Djerassi, C., and Williams, D. H. (1964*b*). *Structure Elucidation of Natural Products*, **1**, **2**, Holden-Day, San Francisco.

Budzikiewicz, H., Djerassi, C., and Williams, D. H. (1967). *Mass Spectrometry of Organic Compounds*, Holden-Day, San Francisco.

Cottrell, T. L., and Walker, I. C. (1966). *Quarterly Reviews*, **20**, 153.

Dewar, M. J. S., and Worley, S. D. (1969). *J. Chem. Phys.* **50**, 654.

Field, F. H., and Franklin, J. L. (1957). *Electron Impact Phenomena*, Academic Press, New York.

Futrell, J. H., Ryan, K., and Siek, L. W. (1965). *J. Chem. Phys.* **43**, 1832.

Hammond, G. S. (1955). *J. Amer. Chem. Soc.* **77**, 334.

Hasted, J. B. (1964). *Physics of Atomic Collisions*, Oxford University Press, London.

Hill, H. C. (1966). *Introduction to Mass Spectrometry*, Heyden, London.

Jennings, K. R. (1965). *J. Chem. Phys.* **43**, 4176.

Kiser, R. W. (1965). *Introduction to Mass Spectrometry and Its Applications*, Prentice-Hall, New Jersey.

Knewstubb, P. F. (1969). *Mass Spectrometry and Ion-Molecule Reactions*, Cambridge University Press, London.

Lederberg, J. (1964). *Compilation of Molecular Formulas for Mass Spectrometry*, Holden-Day, San Francisco.

McLafferty, F. W. (1963). *Mass Spectral Correlations*, American Chemical Society, Washington D.C.

McLafferty, F. W. (1966). *Interpretation of Mass Spectra*, Benjamin, New York.

Mass Spectrometry Data Centre, A.W.R.E., Aldermaston, England.

Massey, H. S. W., and Burhop, E. H. S. (1952). *Electronic and Ionic Impact Phenomena*, Oxford University Press, London.

Mass Spectrometry. (1963). Ed. C. A. McDowell, McGraw-Hill, New York.

Mass Spectrometry. (1965). Ed. R. I. Reed, Academic Press, London.

Modern Aspects of Mass Spectrometry. (1968). Ed. R. I. Reed, Plenum Press, New York.

Reed, R. I. (1962). *Ion Production by Electron Impact*, Academic Press, New York.

Reed, R. I. (1966). *Applications of Mass Spectrometry to Organic Chemistry*, Academic Press, New York.

Rosenstock, H. M., Wallenstein, M. B., Wahrhaftig, A. L., and Eyring, H. (1952). *Proc. Natl. Acad. Sci., U.S.* **38**, 667.

Vestal, M., Wahrhaftig, A. L., and Johnston, W. H. (1962). *Theoretical Studies in Basic Radiation Chemistry*, Aeronautical Res. Lab. Report, 62–426.

Index

acids 82, 84, 85
accurate mass measurement, manual 14, 138
  by computers 14, 126, 138, 139–40
alcohols 72–6, 78, 79, 89, 116, 120
aldehydes 53, 82–4, 90
alkanes 41–2, 65–7, 78, 81, 103, 104
alkenes 67–71, 90.
amides 82, 84, 85, 87
amines, aliphatic 72–5, 77–9, 82, 89, 105
  aromatic 93–6, 120
analysis of mixtures 153–6
apparatus 18–25
appearance potentials 40, 146–9
aromatics, 53, 71, 91–102, 119
  Hammett correlations 97–9
  interaction of substituents 95–7, 100–1
  order of fragmentation of substituents in 96
  ortho-effect in 95

bond dissociation energies, of ions 74, 78, 81, 89, 148–9
  experimental 148–9

characteristic ions 47–9
charge-localization theory 43–5
chemical-ionization 21, 35–6, 163–6
computers, use of in mass spectrometry 46, 55, 126, 139–41
cyanides 78, 82, 94

data acquisition by computers 139–41
defocussing 131–2
derivatives, preparation of, for mass spectrometry 157
directed fragmentation 70, 156–7
double focussing mass spectrometers 22

element maps 55
elemental compositions of ions 12, 51–2
electron-impact ionization 20, 33, 57
equations of motion of ions 22–3
esters 82, 84, 85, 87–9, 93, 121
ethers, aliphatic 72, 73, 76, 78
  aromatic 93, 94, 95
excess of energy in ions, effect on mass spectrum of 1–2, 26–36, 56–63

field-ionization 20, 33, 56, 167–8
field desorption 168
formation of molecular ions 1–2, 19–21, 26–36, 56–63, 163–8
fragmentation, of molecular ions, by simple cleavage 65–103
    by rearrangement with cleavage 107–11
  of fragment ions 103–6

gas chromatographic–mass spectrometric methods 141–6

halides 72, 73, 78, 79, 82, 89, 91, 93, 94, 105, 117
Hammett σ-type correlations 97–9
heteroaromatic compounds 100–2, 119

inlet systems 18–19
ion cyclotron resonance spectroscopy 25, 157–63
ion kinetic energy spectroscopy 132
ion–molecule reactions 159, 161
ions, accurate masses of 14, 126, 138–40
  elemental compositions of 12, 51–2
  formation of, by chemical ionization 21, 35–6, 163–6
    by electrons 20, 33, 57
    by field ionization 20, 33, 56, 167–8
    by photons 20, 34, 56
  metastable 12, 129–38
  multiply-charged 6–7
  quasi-molecular 165

ions, accurate masses of (contd.)
separation of 21–4
ion sources 19–21
ionization potentials 57, 146
isotope abundance patterns 8, 10, 11, 51, 168–70
isotopes, accurate masses of 12
natural abundances of 9, 149
use in labelling experiments 149–53

ketones 82–4, 117

labelling with isotopes 149–53

mass measurement, accurate 14, 126, 138–40
routine 21–5
mass spectra, appearance of 15–17
classification of 46–9
examination of 49–56
formation of 1–6
modification of 56–63
normalization of 2–5, 114–16
of acids 82, 84, 85
alcohols 72–6, 78, 79, 89, 116, 120
aldehydes 53, 82–4, 90
alkanes 41–2, 65–7, 78, 81, 103, 104
alkenes 67–71, 90
amides 82, 84, 85, 87
amines 72–5, 77–9, 82, 89, 93–5, 105, 120
aromatics 53, 71, 91–102, 119
cyanides 78, 82, 94
esters 82, 84, 85, 87–9, 93, 121
ethers 72, 73, 76, 78, 93, 94
halides 72, 73, 78, 79, 82, 89, 91, 93, 94, 105, 117
ketones 82–4, 117
nitro-compounds 93–4
organometallics 72, 74, 85, 102, 105, 123
peptides 62, 85–6, 125–8
phenols 93–4
sulphones 82
sulphoxides 82, 87, 122
thiols 72, 73, 78, 79, 89
metastable ions, 12, 129–38
formation with release of internal energy 135
shapes of 135, 137
mixtures, analysis of 153–6
molecular separators, 141–5
Llewellyn 144
Ryhage 142
variable 145
Watson-Biemann 143
multiply-charged ions 6–7

nitro-compounds 93–4
normalization of mass spectra 2–5, 114–16
normalized abundance tables 2–5

organometallics 72, 74, 85, 102, 105, 123
ortho-effect 95

peptides 62, 85–6, 125–8
percentage of total ion current 4–5
phenols 93–4
photo-ionization 20, 34, 56
product stability theory 39–42

quadrupole mass filter 25
qualitative theories of fragmentation 39–45
quasi-equilibrium theory, 36–8
simplified 37, 38, 77–8, 110–11
quasi-molecular ions 165

reactant gas ions 163–6
rearrangement accompanying fragmentation 107–14
recording of mass spectra, 24–5
computers 25
charts 24
photographic plates 25
Resolution 14–15

separation of ions 21–4
single-focussing mass spectrometers 23
structure elucidation, examples of 112–28
sulphones 82
sulphoxides 82, 87, 122

temperature, effect on mass spectra 60–3
theory of mass spectrometry, qualitative aspects 39–45
quantitative aspects 36–8, 77–8, 110–11
thiols 72, 73, 78, 79, 89
time-of-flight mass spectrometer 25

volatility of samples 156–7